THE
ELSEWHERE
IS
BLACK

THE
ELSEWHERE
IS
BLACK

ECOLOGICAL VIOLENCE
& IMPROVISED LIFE

MARISA SOLOMON

Duke University Press   *Durham and London*   2025

Designed by Matthew Tauch
Typeset in Alegreya by Westchester Publishing Services

Library of Congress Cataloging-in-Publication Data
Names: Solomon, Marisa, [date] author.
Title: The elsewhere is black : ecological violence and
improvised life / Marisa Solomon.
Description: Durham : Duke University Press, 2025. |
Includes bibliographical references and index.
Identifiers: LCCN 2024060392 (print)
LCCN 2024060393 (ebook)
ISBN 9781478032465 (paperback)
ISBN 9781478029137 (hardcover)
ISBN 9781478061304 (ebook)
Subjects: LCSH: Environmental racism—Atlantic Coast (U.S.) |
Environmental justice—Atlantic Coast (U.S.) | Refuse and refuse
disposal—Atlantic Coast (U.S.) | Environmental health—Atlantic
Coast (U.S.) | Minorities—Health aspects—Atlantic Coast (U.S.) |
Minorities—Health and hygiene—Atlantic Coast (U.S.) | Political
ecology—Atlantic Coast (U.S.) | United States—Race relations—
Economic aspects.
Classification: LCC GE240.U6 S656 2025 (print) |LCC GE240.U6
(ebook) | DDC 304.2/808996073—dc23/eng/20250512
LC record available at https://lccn.loc.gov/2024060392
LC ebook record available at https://lccn.loc.gov/2024060393

Cover art: Curtis Cuffie, sculptures, c. 1994–1996.
Photo by Katy Abel.

Publication of this book is supported by Duke University
Press's Scholars of Color First Book Fund.

To all my irreverent and disreputably gendered

environmental teachers on the street,

this book is for you.

Contents

Acknowledgments

I have always dreamed of being a writer. But as a Black working-class immigrant kid with cancer, I missed critical years of my education—namely, the years when students begin to learn grammar. Those rules have never made sense to me, and it has often made my desire to write feel like a painful compulsion. And so I've spent many years writing in and with shame. It's difficult to describe, even here, how the impossibility of Black childhood, Black girlhood, illness, and poverty constrains what you think you're capable of. It's even harder to describe how it casts a shadow on what you feel you can ask for as a (Black) adult. Despite what feels like years of hiding, I have also benefited from those (very brave) people who were willing to knock cautiously at the door of my existential dread, embrace me, and open up worlds of possibility that my young self never thought possible. Without them, this book does not exist.

On my first assignment as a graduate student at The New School, Miriam Ticktin, my long-term mentor and now colleague and friend, wrote, "See me after class." Those words nearly obliterated me. Avoiding eye contact with her, I skirted the more well-read students lining up to talk to her, hoping to make it to the elevator, out the door, and onto the street before anyone could see me cry. But before I could make it down the hallway, she caught my eye and waved me into her office. What I thought would be a meeting about how I needed to work on my writing (though we would have those too) turned into a first-of-many conversations about labor, value, capitalism, and race with a powerful anticolonial feminist scholar, activist, and pedagogue. Before I knew who Marx was, Miriam knew I was a Marxist; her confidence would lead me to the Black materialist questions about racial capitalism at the heart of this book.

My interest in *matter*—fraught with what at the time was called "the new materialism"—was guided by Hugh Raffles's commitment to prose and storytelling. His work pushed me to ask, What does matter *do*? I am grateful for the not-always-gentle hand with which he approached my words, for the way he pushed them and me to *do more*, and for offering me the building blocks to approach description as an argument itself. I also want to thank Rachel Heiman, without whom I might never have discovered that I wanted to be a teacher. Presenting in her The Global Middle Classes course was a pivotal moment in my education: It was the first moment I realized that even if I couldn't write, maybe I could *teach*.

The chronic panic stalking me through graduate school meant that I almost dropped out more times than I can count. If not for Sharika Thiranagama's encouragement and support, I wouldn't have stayed in graduate school long enough to find out if I could teach or write. I continue to strive toward her ability to think deeply and widely. Finally, though no less impactful, is the mentorship of Jacqueline Nassy Brown, not because her Anthropology of Race, Space, and Place class was where I met the love of my life but because her enduring commitment to anthropology's possibilities remains a guiding ethic in my work. Storying the marginalized histories, as well as the minor stories of places shaped by chronic dispossession, motivates my ongoing, if perhaps wayward, relationship with (a sometimes more imagined than practiced) anthropology, unmoored from its colonial imperatives.

But mentors don't only come in the form of professors. I am lucky enough to have friends who ground and elevate me. To Katie Detwiler, who has one of the sharpest and most expansive minds I've ever met: I find myself awestruck by the way you turn ideas, concepts, phrases, and epistemological tools around to produce new analyses, lines of thinking, and potential for expansive relations. From spirit animal card readings at our neighborhood bar to late night phone calls in the field, our relationship has taught me how to exercise intellectual muscles I didn't know I had, that fun and rigor are not mutually exclusive, and that theory is not the opposite of practice. To Jais Brohinksy, I'm eternally grateful for a friendship that holds open space for me to confront the best and worst of me. Your relentless questioning of why we live the way we live continues to teach me what you find when you dwell in the contradictions of capitalist ways of life, and I never want to write or live without it. Christopher Paul Harris, kin and comrade, friend and fam, thank you for reminding me that Black joy is made in communion. To Jacqueline Katanesza, I am grateful for your love, friendship, and support—from reading early drafts to listening to me teeter on the edge of sanity while writing. To Manijeh Moradian, thank you for holding me down when my self-doubt has me ready to fly the coop; with a flame as radical as your love, you cleave open a place (in the academy) for me to decide when I'm ready to ground. To Rhea Rhaman, from whom I learn bravery: I'm in awe of you. Beyond your ability to scale a rock, determination to learn to surf in order to face your fear of water, and relentless insistence on facing yourself, your friendship has been a buoy reminding me that sometimes what I need to say requires diving deep to reach something submerged beneath the surface.

In my first tenure-track job, I met some special people who made it possible for me to write this book in the best and most inconvenient way possible: the way I wanted to. Thank you to Donika Kelly, for shifting my perspective; to Marcus Johnson, for gathering us dorky Black folks together to just *be*; and to Carolle Charles, for offering your grace and wisdom in the face of hostility. I learned so much from Angie Beeman, my colleague, friend, and mentor, who reminded me that fighting for each other is how we fight for ourselves. A special shout-out to Angie's children Hope and Justice, whose profound sense of self in the face of obstacles is an honor to witness. This book would not have happened without the love and friendship of my Black feminist co-conspirator Erica Richardson. What began as discovering a shared love of sci-fi and horror films became the pivotal intellectual romance of my life. Both in the classroom—including our co-taught Black Archaeologies class—and in the living rooms we make possible, with you I've learned what Black feminist study feels like.

As anyone who has written a book knows, each word, each sentence, each period, is the labor of many conversations and iterations, much practice and feedback. To my editor, Elizabeth Ault, who has been more generous with her time and my writing than most, and to her assistant, Benjamin Kossack, who had the hard job of getting me to let go of the writing: Thank you for believing in this book before it was a book. I also had the true honor of sitting in a room with Neferti Tadiar, Sarah Haley, Terrion Williamson, and Sarah Cervenak for a manuscript workshop. While I was starstruck, the group's tender rigor and attention to Black feminist theory pushed this manuscript into new domains. I also want to thank Vanessa Agard-Jones for inviting me into the Black Atlantic Ecologies (BAE) working group at Columbia's Center for the Study of Social Difference. Her work on colonialism, materiality, toxicity, and Black life set the stage for my work. I also want to thank the members of BAE—Aimee Meredith Cox, Jayna Brown, Amber Jamila Musser, Julie Livingston, Sonya Posmentier, and Anna Arabindan-Kesson—for their support during the early stages of this work, and Chazelle Rhoden and Alyssa A. L. James for facilitating our gathering, a difficult task during those early years of the COVID-19 pandemic. I also want to thank Cajetan Iheka for giving me the opportunity to share my work at the Black Environmentalisms Symposium at Yale in 2022 and Robyn d'Avignon for facilitating our introduction. The political and intellectual generosity of J. T. Roane, Rasheed Tazudeen, and Elleza Kelley deeply inform this work. I also want to thank the collaborative group of scholars at NYU who come together under the banner of Discard Studies. Thank you to Rosalind C. Fredericks for giving me the opportunity to share my work with this crew, which includes Elana Resnick, Amy Zhang, Robin Nagle, Joshua Reno, and Sophia Stamatopoulou-Robbins. A special shout-out to Max Liboiron, who not only began the Discard Studies blog but early in their career launched the Object Ethnography Project, in which Vincent Lai and I had the pleasure of taking part. Through their creativity, resistance to writing any other way than in their own voice, and commitment to radical decolonial praxis, Max remains a breath of fresh air. I've been learning from them since we met.

Along the way, I've benefited from some truly special individuals who have a talent for cultivating spaces of camaraderie against the individualist foreclosures of the university. First, thank you to the Studying Is Not Enough reading group, Eae and Vani in particular, who were unjustly fired from Barnard College and whose bravery and political clarity are an inspiration. Second, I want to thank Rebecca Jordan-Young and Elizabeth Bernstein for one such space, which goes under the name Recovery. For the last two years, the conversations among the members have been a feminist homecoming. Thank you to Jackie Orr, Miriam Ticktin, Kerwin Kaye, Amy Zhou, Sam Roberts, Nadja Eisenberg-Guyot, Chloé Samala Faux, Salma Ismaiel, and Gabriel Quintero for your intellectual generosity and for a downright good time. Third, I want to acknowledge how rare a thing it is to find a political home in any department, but in the Women's, Gender and Sexuality Studies Department at Barnard College, I have found just that. To Rebecca Jordan-Young, Manijeh Moradian, Elizabeth Bernstein, Neferti Tadiar, and Janet Jakobsen: Thank you for continuing to teach me what feminist commitments look like. But feminist commitments, *Black* feminist commitments, must also be to oneself. With the help of Celia Naylor and Tami Navarro, I'm learning that you can't make those commitments alone. To Celia, who offered up a piece of wisdom exactly when I needed it most, thank you for the words that have inspired me and will serve as a lifelong lesson in moving toward my peace. And to Tami, thank you for the Black feminist space of doubt, possibility, and accountability. Thank you to Ashley Ngozi Agbasoga and Leniqueca Welcome, a support system through a tumultuous year, both politically and personally. In a world that chokes the breath out of Black women, I am so grateful for the breathing room.

Thank you to all of my students at The New School, Baruch College, Columbia University, and Barnard College, who inspire me to keep writing even when, or perhaps especially when, I think I can't. This year, 2024, in particular, a year of multiple and multiply connected genocides, I've been motivated by some particularly brave and brilliant people who taught me that sometimes students are the teachers and the teachers are the students. May we all work toward being members of the Red Fan Club.

As most Black people know, the Black family is not accurately described by the cis-heteronormative family. With aunties who are community mothers, brothers from other mothers, cousins who are sisters, baby daddies who are absent, fathers who are outside the biological relation, and Black mamas who are mothers, fathers, and sisters all in one, my Black (interracial) family is queer in this expansive sense. I have many parents, and I'm grateful for all of them. But most importantly, I'm grateful to have been raised with an expansive sense of family, which so deeply informs how I think and theorize kin. So to all my parents—Ruth, Bill, Addie, Lloyd, Grandma Florence, my aunt Stephanie, and my dearly departed papa—I'm grateful for the non-normative relations that make me possible.

But also, my family is queer AF. I wrote this book in large part to honor the hustling, labor, and fire of my queer mother, Ruth Murray, whose enduring commitment to learning about the world and herself inspires my own. When she came out when I was sixteen, I gained another mother, Addie Clark, whose quiet yet large love for life's details has taught me to be tender with the world, to cherish relation, and to be in awe with all beings. Both my moms, in different ways, teach me to keep humble wonder in the face of the unthinkable. To my sister, Mariah Murray, whose spiritual journey has taught me that life does not always lead you where you might think, I'm grateful for your steadiness. To my youngest sibling, Hero Garland King, one of my absolute favorite people in the world, I'm grateful for your fabulous queerness, your humor, your sensitive observation, and your brilliance—after all, you coined the term *homotional* to describe us all. To my grandmother, Florence Elliott—whom I include here as an honorary queer and without whom my life would not exist—I am eternally grateful for what you made possible for all of us: a chance to learn how to transform on our own terms. Without you, I would not be who I am. None of us would.

When I fell in love with my partner, my family got even bigger. Their love and support have made not only this book but the person who wants to keep writing possible. To Mark Eisenberg, whose longtime activism and work as a doctor have taught me that we must be advocates of harm reduction in all aspects of our lives, thank you for being a sounding board, a storyteller, a music lover, and a parent who welcomed me without hesitation. To Kris Guyot, who read my dissertation from cover to cover and sent me a handwritten letter steeped in intellectual generosity and curiosity, you've never made me feel anything but a missing and then found part of the family. It is a lifelong gift to be held in the embrace of your Marxist-feminist fire. To Jerzy, neighbor, friend, brother, and comrade, I'm humbled by your wit, your mind, your discipline. I'm grateful for how you always show up for family, and I admire how hard you work, not just on your own politically urgent scholarship but on knowing yourself. Thank you for being my brother and for bearing witness to some of my life's most important moments.

Too many of these words were written during the mass death of people of color. Amid the litany of violences, I spent too many days writing in hospital rooms, fearing that I might lose some of the most important people in my life. I spent my childhood in hospital rooms, with a multitude of anxious kin, already stressed from their multiple jobs, fearful that I might never make it to ten. Then and now, I continue to learn that a life circumscribed by death makes love, friendship, and community the most precious things we have.

I was blessed enough to have found such a precious love at the young age of fourteen with my best friend. Even in high school, when we were traumatized by life and "in love" with the same gay boy, Elana Lopez was the most brilliant, committed, intentional, loving, and ethical person I had ever met. As bell hooks once wrote, "Many of us learn to love not in the context of family or in romantic relationships but rather in the context of friendship." It has been over the more than twenty years of our friendship that I've learned why, as hooks also wrote, friendship is a "threat to patriarchy." Elana, from you I've learned that friendship is a liberatory romance, a commitment to a place where we make each other the antiracist, decolonial feminists that we want to be. From you I've learned that in a world that kills, we must tend to each other, especially because the conditions of injustice that make the everyday violent make it too easy to stop trying to build. With you I've learned that love means embracing change, that building means reflection, and that it is in those evolving reflections that we honor our ongoing histories and figure out how the present, not the future, is where we struggle for liberation.

It's hard to believe that Alex Lopez—the shy, sweet Cuban boy from my fifth grade gym class, who became a teenage bad boy/tortured poet—is now the thoughtful organizer, teacher, farmer, and most admirable father I've ever met. While some things have changed—biting teen boy sarcasm softened into dad jokes—your political fire and commitment to labor politics have not. I'm inspired by how you learn from the ecologies around you, whether that's the power dynamics of unions or land in its complexity and ecological commensality in the context of settler colonialism. Watching Alex and Elana parent is humbling and grounding because parenting is the everyday work of transmitting values with radical love. Together, the two of you have given me more than I could ever thank you for, including the lights of my life: my three beautiful godchildren. I am humbled my Malkah's perseverance and attunement to others; I'm in awe of Ayden's thoughtful mind; I'm in love with Kai's humor and the worlds he builds with his imagination. They are undoubtedly their own brilliant humans, but they are also modeling, living, and becoming people with the values that guide you both.

Elana once described our friendship as guided by the improv principle "Yes, and." I agree, but I would say that improvisation—the way that one changes course, adjusts, moves because they have to and, in so doing, moves others—is the way you live your life. And your improvisations are life giving. To Elana, Alex, Malkah, Ayden, Kai, and the more than humans (Shana, Frankie, Tali, the land in all its complexities, and all the chickens past and present): Thank you, all of you, for a history and a forever of "Yes, ands."

This book is, among many things, an intellectual love letter to the Black and trans romance I've been blissfully tangled up in for the last ten years. Nadja Eisenberg-Guyot, from the margins of the books we share to the marginalia of years-long conversations, I've learned from you that the point of theory is to learn how to love better, to never stop tending to the ground from which we do it, and then to share that ground with as many people as we can. You've taught me that in practice that means learning how to refuse, making generosity a political weapon, and being steadfast in our desire for other people's freedom—in the room, on the street, at the dinner table, in front of peers—and modeling it for our students. While books are wonderful, they are what brought us together; as an organizer, you remind me that people and their capacity for transformation are better. I strive for your political clarity, your capacity for joy, and the way you never mistake the site of struggle for the site of liberation—something that the university seductively confuses. I'm forever grateful for the tenderness with which you always approach my words, both on the page and in life. Without you, this book could have never become what it has for me: a commitment to finding abundance, even when the world is trying to kill us both.

This project was made possible with funding from the National Science Foundation Graduate Research Fellowship, the Wenner-Gren Dissertation Fieldwork Grant, the Graduate Institute for Design, Ethnography and Social Thought (GIDEST) at The New School for Social Research, and the CUNY Faculty Fellowship Publication Program.

Introduction

INTRODUCING THE ELSEWHERE

It is a narrative written from nowhere, from the nowhere of
the ghetto and the nowhere of utopia.
— HARTMAN, *WAYWARD LIVES, BEAUTIFUL EXPERIMENTS*

We are the guerrilla poems written on walls, purveyors of a
billion dangerous meanings of life.
— GUMBS, "KEYWORD: MOTHERING"

In my family, we tell a story about me; I tell it too. And in many ways, it is
where *The Elsewhere Is Black* begins. When I was young, I used to eat books,
my mother's in particular. She collected them everywhere. She would buy
trashy novels from the grocery store and pick through every abandoned
box of books she encountered on the sidewalk, sometimes taking the
whole box home. She would swipe books from the doctor's office and
adopt abandoned novels from street corners. She happily accepted books
from friends, no matter what they were about, and when we would venture to the Salvation Army, she would make a beeline to the ten-cent book
bins and buy as many as she could afford. Even if they were torn, even if
I had gnawed at the edges, she would still make her way through them—
sometimes crafting imagined line endings at the eroded margins, sometimes reading them as if that was the way they were supposed to be.

No matter how trashy, books have always been the material with which
my mother set herself aflight from the narrowed possibilities of racial
and sexual condemnation. At fifteen, she didn't know that women in rural
Tobago were being inundated by pesticides, so she couldn't have imagined
that those pesticides, themselves bound up in long histories of colonial violence, would be part of the ecological conditions of my birth.[1] She was also
too young to notice that the passing of Trinidad and Tobago's 1986 Sexual

Offences Act shaped her struggle to name her queer desires, particularly because she was a pregnant fifteen-year-old "dougla."[2] What she *did* know was that reading was like breathing, and it was a practice she carried with her when she was reunited with my grandmother, who had been working tirelessly as a domestic worker in Brooklyn, trying to make it in America and bring her children over too. But my story, and my mother's and grandmother's, is like most Black immigrant stories. America might have proved different from the racial-sexual economies that constrained Black women on the twenty-six-mile-long island of Tobago where I was born, but the geographies of racial capitalism in the United States did not exactly offer more room for a Black teenaged mother (or her daughter) to *be*.

It was the late 1980s when my mother and I came to the United States, a time when the privatization of city services was on the rise and the scope of the anti-Black carceral state was changing. The legacy of redlining, of blockbusting, of white flight, and of disinvestment were well-entrenched in cities with large Black populations. Hartford, Connecticut, where we ended up, was no different. By the 1990s, the conflation of broken windows and trash on the streets with Black criminality was being consumed in the form of shows like *Cops*, while hip-hop was understood to be a hotbed of deviant Black working-class aesthetics pejoratively labeled "street." Black women's aesthetics in particular symbolized the trash coming from the hood. The "welfare queen" and her many names gave national grammars to the gendered equation of Black degeneration, and her progeny were being thrown in jail at younger and younger ages.[3] The school-to-prison pipeline was hardening as the welfare state was disappearing. And under the guise of protecting (white) women from the "superpredators" coming from the hood, the ghetto had crystallized in the white supremacist imagination as a tangle of pathological relationships to space, an "over there" problem of Black genders and their impropriety, a ghetto that is simultaneously elsewhere and somehow not of this world at all.

The ghetto, as the late Steven Gregory theorized, is a term that conceals more than it reveals, a term that obscures the material relations of depletion, neglect, or what Katherine McKittrick might name as the way geography becomes a form of racial-sexual condemnation.[4] Constitutively, the processes of material and ideological condemnation that produce the cartographic struggles of Black life hinge on the contradictory premise that Blackness in the United States is essentially *unecological*. The term *ghetto*, then, also conceals the eco-logics of property, whereby the protection of property and white life purports to revere (while simultaneously destroying) a "pristine" nature that "seems innocent of black history."[5] The condemnation of poor people of color's relationships to their (gendered) bodies, to objects, to place, and to environments casts Black, especially poor, urban life outside nature-social relations. Yet, seemingly paradoxically, the condemnation of "the Negro" through their relationship to (or *as*) wild is used to construct Blackness as animality in order to *naturalize* exploitation and criminalization.[6]

Whether it be the inner city, the suburbs, or a rural area, Black people are presumed to be both agents of disorder and degeneration (unecological) and a people overdetermined by "broken windows," "hot climates," and "savage jungles." Most clearly operative in policing tactics, the long-debunked-but-still-practiced criminological theory of broken windows specifically targets "signs of decay" as a tactic to deter "criminals" from antisocial behavior violating order and lawfulness.[7] Yet as many critics have pointed out, what cops (who are often property owners themselves) actually reinforce, if not produce, are racist perceptions of Black people as inherent hazards to property.[8] My point here is not to survey criminological theory so much as it is to point out the lay environmental theory that lies at its heart: Blackness is a pathological, if not criminal, excess that disorders environments, and the environments of Blackness are a pathological hazard of criminal excessiveness. Black people are contaminated—a state that has already come to pass—and contaminating—a threat to sanitized white life.

Take my eating habits. Along with my mother's books, I also ate paint. And when I was three or maybe four years old, I was required to take an IQ test (not a blood-lead-level test) to prove I was "delinquent." After all, as one of my teachers remarked, I was eating books. Eating paint was just more evidence of an inability to engage respectably with objects, evincing my young propensity for disrespecting the property of others. The lead levels in the two-bedroom house where we lived with eight other family members were never tested, but we didn't need a test to tell us the levels were high. We knew we lived in the ghetto, and that was enough to know some things just weren't right. Sometimes we speculate about the potential link between lead and the diagnosis of leukemia that came when I was five, but we do so cautiously. The elision of my body and our environment would be a trap for Black women, a story well-worn in the annals of history. It is the story of the pathological Black mother who fails to rear Black children in acceptable or healthful environments. Under her many names—welfare queen, hood rat, ghetto trash—she produces children (no matter their genders) who inherit her propensity to degrade the environment.

In this equation, the ecologies of Blackened places are not only punishing for Black women; the landscapes of their environments also become interchangeable with Blackness itself. The particulate matter caused by a "green" waste-to-energy plant; lead in the patch of grass behind the projects, which were built over a lead-soaked field; unused lots that are actually Superfund sites, with unknown measurements of arsenic and unnamed chemicals, become the often-invisible backdrop of places condemned by the anti-Black state to the condition of blight. The irony is that if Blackness is interchangeable with a broken window, a pothole, or any other sign of state neglect, then Blackness is ecologically useful for racial capitalism. Blackness marks an elsewhere to which risk can always be consigned, and it marks such material conclusions because Blackness is not *ecological* at all.

TOXIC CAPTURE, UNECOLOGIC, AND PROPERTY AS A GENRE OF LIFE

The "unecologic"—riffing off McKittrick's use of the Black "ungeographic"—points us to how the priority of property casts Blackness as excessive to and outside of the environment.[9] Forged in the architecture of the settler-master's plantation, colonial expropriation of Indigenous homelands wielded enslaved bodies as an environmental weapon. And it was through this environmental process that the settler made himself into what Tiffany King calls the "conquistador-settler."[10] Through (il)legal documents, he created his own property in land and people. He became his own cartographer, mapping *over* Indigenous territories, claiming theft as the law.[11] And with new maps, he installed a plantocracy, making himself the author of antagonistically organized difference that sought to obliterate native claims of kinship with land, naturalize the destruction of delicate ecologies, criminalize Black life off the plantation, and ensure his own genre of habitation (property).[12] In the entwined genocide and ecocide of his own propertied becoming, the conquistador became the steward of death. As Haunani-Kay Trask succinctly said of the United States, "Colonization was the historical process, and genocide the official policy."[13]

The environmental devastation of Black and Indigenous dispossession is fundamental to what underpins settler self and ecological becoming. As Eve Tuck and K. Wayne Yang argue, a violent racial triad (settler-native-slave) supports the settler colonial formation: a native to be disappeared and a slave whose labor was central to value production but whose presence and personhood were considered an excess to (stolen) land and/or environment.[14] Materialized as "excess" by the geographic institutions and environmental policies that uphold (settler) whiteness as property, the ongoing production of the settler-native-slave triad makes Blackness maimable, displaceable, and killable, ever held captive by the need for land and/as a "sink for pollution."[15] The violences of (white) property emanate from the epistemic, ontological, and material force of the social relations of ownership. Yet the risks of this propertied habitation accumulate elsewhere, subjecting Black life to toxic capture.

Throughout this book, I emphasize that anti-Blackness is, among many things, an environmental strategy. If the unecologic (or, rather, the racist eco-logics of property) casts Black (especially wageless) life outside of nature, one of settler racial capitalism's uses for the Black body is Blackening place to make way for waste disposal and toxic accumulation. Unceded lands "have been forcibly made part of the US [waste] infrastructure," and stolen Black bodies are inexhaustibly used to absorb, and plan for, the places of environmental risk.[16] I call this ever-accumulating and compounding condition of risk "toxic capture," to expand our thinking about environmental racism to include the way that property is a genre of living (by killing). Though some of this book takes place near sites of waste disposal (landfills, transfer stations), I argue that property itself is inherently environmentally destructive, including through its production of waste management's mundane yet disavowed toxic conditions.[17] Despite the reverence the conquistador human has for how he constructs and bounds nature, property is not ecologically sound. The management of waste relies on its placement. And planning the places of waste requires the calculation of risk to bodies, to water (tables), to marine life, to soil, to air—a colonial science that, even if not always immediately obvious, racializes land "use" to determine "the threshold of harm."[18] While environmental regulations attempt to manage the risk of pollution to human and more-than-human life, they do so by disavowing a (settler) colonial capitalist system that uses the Black body to expropriate unceded native land and constructs "Blackness" as a "human and environmental hazard."[19]

Blackness is environmentally useful for capital, by which I mean productive through its condemnation of Black people as an environmental problem. Toxicity follows capture wherever Black life is made quintessentially disposable, and it is the accrued but disavowed consequences elsewhere to property as a genre of habitation. Materially, this means that our collective and more-than-human environments are enrolled in the project of domination, including the project to cast out (or punish). But casting out does not actually make material (or ecological) relations disappear; instead, it changes them and often with dramatic consequences that threaten if not reveal the instability of property and its attendant ways of living. As a consequence of the unecologic, waste/risk/toxicity is what property produces, and capture is an ever-available method to protect a propertied way of life. Thus, toxic capture describes a racialized material condition; it is an ever-accumulating reality and a condition produced by (settler) colonial racial capitalism's most destructive fantasy: that property and its genre of living (or killing) are (or ever were) sustainable.

Toxic capture appears in different ways across the places and times of this text, sometimes next to the acute sites of waste management, sometimes as the seizure of Black life by eminent domain. My hope is that naming toxic capture as a condition makes it imperative to see those discarded and made discardable by it as environmental theorists. Their ways of talking, telling, improvising with, speculating about, and storying matter are instructive ecological ethics for living with/in and against a material world, to reprise Carolyn Finney, that is seemingly innocent of Black material histories. It is precisely the conditions of toxic capture that make the Black ecological struggle also a struggle with the terms of ecology (as pristine, clean, and white).

Surrounded by toxic capture, I, too, struggle to make sense of my childhood with the dominant terms of ecology, terms that do not address the contradictory roles Blackness is forced to bear vis-à-vis nature and environment. What if those contradictions did alter my hunger? What if bearing down on my stomach was the need for nourishment to be more than food? What if my hunger (for books, for paint) charted alternative forms of stewardship through Blackened depravation? If reading offered my mother a way to move, then perhaps eating books was a way for me to be with her in ecological struggle, a way to improvise with matter's "contingency and possibility" instead of property's "necessity and determinacy?"[20] What if, in the midst of being hungry, consuming indigestible paper, PCBs in ink, bacterium, dust, and glue, I was experimenting with the terms and relations of my environment? The Blackening of place is part of what M. Murphy calls the conditions of altered life, in which "life already altered . . . is also life open to alteration."[21] Living the acute "contradictions of existing in worlds that demand chemical exposures as the condition for eating, drinking, breathing," what if eating books and paint was a way for me to improvise within the violences of being discarded?[22] What if it was a practice that exposed and indexed how my hunger was the product of many things (poverty, loss, racial and sexual condemnation)? And the things I was hungry for (my mother, flight, what reading represented, nourishment, pleasure) were as multiple as the world's violences? Perhaps in a world that wastes life, I was rendering the world differently with waste.

A professor who used to eat books is not an ecological story that fits neatly within the history of environmentalism, conversations about climate crisis, or more recent moves toward zero-waste futures. It is not a story of renewable energy (though it is perhaps a story of how Blackness is ever renewed as energy for capital), and it is not a story that coheres under the rubric of green techno-optimism that we might note in climate accords. In fact, it's a story that challenges the optimism of zero-waste futures, and tasks us with including trash in our understanding of the Black present, a vision that insists on prioritizing (the poor, the houseless, the wageless) Black presence. That my family continues to tell the story of how I ate books, and that I tell it here, too, is a way to *not* elide the conditions of our life by rendering them only abject, or as the failure of Black family formation. Instead, we tell a story that neither pathologizes nor cleans up; we tell stories about the other ways that we *are*, and in so doing, we tell stories of other ways of being.[23]

Along with the toxicity I undoubtedly ingested, I consumed other ways of knowing the vexed political object upon which this book hinges: waste. The unsanctioned practices of reuse deployed by poor and working-class people are often criminalized precisely because they have the potential to rearrange the order of value, or because they capture something that exceeds capitalism's clutches. For example, in 2016, when the New York Department of Sanitation (DSNY) realized that the city's curbside waste diversion rates were nearly half that of the national average (16 percent and 34.4 percent, respectively), the "theft" of recyclables became an environmental issue. With Mayor Bill de Blasio's zero waste by 2030 campaign, the DSNY was under the direction of the mayor's office to increase rates to 25 percent. Although urban scavengers already do the labor of diversion, Vito Turso, spokesperson for the DSNY, told the *New York Times*, "The city's got goals, and the only way we know how to meet those goals is if we have control over the commodity"—that is, recyclables.[24] Despite the bottle bills (passed in New York City in 2008), which allow for the scavenging of certain recyclables (with some restrictions, such as prohibiting the use of vehicles), the criminalization of scavenging points to both the value of waste to capitalism's racist order and the fugitive politics of discarded living.

Making life among discards is elsewhere to an environmentalism that hinges on capitalism's attempts to save the future for itself by "controlling the commodity." Zero-waste campaigns, diversion rates, and recycling "theft" are among the many iterations of capitalism's desire to clean up places and people for its own accumulation. However, to name the possibility of a different genre of earthly knowledge, a different orientation to the matter of living under intersecting forms of domination, I join L. Horton-Stalling's stated desire to embrace "the dirty," which I extend to the *trashy*. "The dirty," Horton-Stallings argues, "in its finest and filthiest iterations . . . exists as the simultaneous place and practice of intersectional politics, critiques of moral authority, and the development of regional aesthetic philosophies whose purpose is dismantling and reinventing southern public spheres largely erected out of the sexual economy of slavery and sustained by settler colonialism."[25] The way that *dirty* signals both a condition (relative to cleanliness) and an aesthetic racialized sexual politics is critical to understanding the intervention that dirty makes. I seek to wrench open *trash*, too. *Trashy* is as much a state of capitalism as it is an aesthetic practice, or in Jillian Hernandez's term, an aesthetic of excess.[26] My hope is that straining to see the relationship between trash and *trashy* offers a different way to think about waste, trash, and the discarded as material sutured to dispossessed life. In addition, I hope that this prompts a rethinking of the meanings of *waste* and its accrued (social/moral) imperatives.

KNOWING WASTE

Marco Armiero argues that "waste is not a thing to be placed somewhere"; rather, he urges us to shift our attention to "a set of wasting relationships."[27] This shift to waste-producing relations reveals waste-facility siting to be far "more than a matter of miles and zip codes." While violent spatialities are key to a waste management system that prioritizes the health of white property over all other things in the United States, waste is also a *relation* that "*produces* the targeted community" rather than merely an unfortunate consequence of the science of land use and its spatialization of discriminate *zones*.[28] Thus, in agreement with Armiero's critique, this book does not approach waste as a metaphor for dispossession but as dispossession's material reality.

The discarded pile up, onto, and perhaps *into* one another, producing a toxic tethering, wherein waste becomes a condition of dispossession.

Waste matter and racialized bodies stick to one another within racial capitalism's geographies; however, the spatial violence that subtends this mode of discarding can't always predictably "incorporate" its own excesses. Against narratives that presume that the dispossessed either are a waste or waste their lives, "subaltern people have gotten organized, sometimes openly clashing with the forces of wasting . . . often struggling to substitute wasting relationships with commoning."[29]

Recent work on commoning proves important to the way I theorize the materiality of dispossession. For example, Miriam Ticktin's work points us to the ways that poor, racialized, and gender nonconforming peoples have generated creative ways to survive by producing new forms of commons, despite racial capitalism's enclosures.[30] "Commoning practices," Armiero argues, are a form of sabotage that interrupt the social relations that waste.[31] The feminist injunction here, counter to romanticized and patriarchal visions of "the commons," is to see practices of commoning as part of the mutual-aid work required to produce alternative futures. In a similar vein, J. T. Roane historicizes the Black commons as a practice and a place of freedom forged on the plots of the plantation.[32] Roane's work is particularly instructive here, as he traces the seemingly unruly material practices of slaves (in the stolen seed and the stolen time of plotting) as insurgent knowledge production that allowed the enslaved to use their familiarity with the environments of the plantation to craft methods of resistance.[33] Sometimes this resistance was in the form of escape, in which "the swamp and the wilderness" served as shelter.[34] Sometimes it manifested in resistance to the settler-planter economy in which the slave was a tool burnished within the economies of genocide and the disposing of land.[35] Regardless of how the knowledge was acquired, it was used deceptively, thoughtfully, agentively, to forge freedom with unruly matter.

It is in this spirit that I approach the materiality of dispossession, or waste as dispossession's attendant condition. Ethnographically speaking, this means I prioritize the knowledge of the dispossessed, whose object practices and material relations—including drug use, cigarette smoking, and aesthetic cultures—are often pathologized, if not criminalized, by a carceral imagination that approaches discarded life as itself an environmental hazard. Instead of solely theorizing the disposability of "redundant populations," this book takes knowing waste through dispossession seriously as a potential site of insurgent ecological knowledge.[36]

Drawing on ethnographic research in the Hampton Roads (or Tidewater, depending on who you ask) region of Southeastern Virginia—currently experiencing transformation across the cities of Norfolk, Portsmouth, and Suffolk—and the gentrifying neighborhood of Bed-Stuy, in Brooklyn, New York, this book approaches the material practices of Black dispossessed persons as ecological knowledge. Though this project began with scavenging on the streets of Bed-Stuy with houseless Black men, I also found my way South, attempting to trace the circuitous logics that make *Black* and *trash* redundant.[37] In Virginia, where landfills rearticulate the plantation's resilient relations, I found myself learning the most about trash, trashiness, and Blackness from sex workers, who proudly referred to themselves as "trashy hookers."[38] Through the stories that scavengers and sex workers tell, to the paths of critique that their movements through space chart, this book argues that we need to pay closer attention to the materiality of Black life as the experimental ground for new forms of ecological ethics.

In order to launch dispossessed Black living as fecund with abundant ecological politics, we first need an important corrective. Kathleen Millar points out that waste is a metaphor that reinscribes capital and the state. She argues that "those whose labor is *not needed* by capital" are those described as "waste," "surplus," or "superfluous" to the state.[39] However, as waste historian Susan Strasser's work makes clear, until the twentieth century in the United States, laboring waste *was* needed by capital, though it was primarily unwaged labor, bracketed off from the "productive" economic sphere altogether.[40] From street gleaners to rag pickers to the fat rendering, soap making, and mattress stuffing practices of the domestic arts, laboring waste was socially *re*productive.

As I interpret Millar's argument, the reinscription of waged labor as the only way to understand value and political economy is a classist—not to mention sexist, which is always racial—epistemic problem in the scholarship on waste and an ideological undercurrent in the response to it. To put it differently: We presume to know why people dig through the trash, and thus we presume that we already know the story. Not only does this make classism a problem for how scholars and ethnographers "know," but it also turns dispossessed people's relationship to waste into something that is *only* a response to economic scarcity. This assumption is a form of epistemic violence, in which the only thing that dispossessed people can do (on the page) is survive. While employment and making a living "is . . . a central dimension of *a form of living*, it is not synonymous with . . . the pursuit of a specific mode of inhabiting the world."[41]

Making a living, particularly when you are barely apprehended as human, isn't the only way to live in an anti-Black world. And in the lives of the wageless sex workers and scavengers I learned from who live on the street, the value(s) of humanism and its definitions of work do not guide living. This resonates with one of the central arguments in Neferti Tadiar's *Remaindered Life*, which centers the violence of humanization as a "value-constitutive activity" that presumes to know the labor of "a life worth living."[42] Building off Tadiar's and Millar's critiques of how stories of waste can become austere, this book, too, seeks to "break open this reduction, allowing for diversity of productive actions that do not fit easily into capitalist categories of labor and notions of work."[43] In fact, in this book, the people who tell stories with and about waste, as well as labor on it, revel in other ways to *be* amid a dispossessing world, including through the proliferation of genres—of aesthetics, of gender, of the human—that build socioecological relations from the colonial detritus of our present, deeply haunted by its pasts.[44]

Self-proclaimed garbologists William Rathje and Cullen Murphy ask, "If our garbage, in the eyes of the future, is destined to hold a key to the past, then surely it already holds a key to the present."[45] This provocation to think of garbage as a socio-ecological record presents an important companion to Myra J. Hird's argument that knowing waste is an "ironic testimonial to our desire to forget."[46] Hird argues that waste infrastructure doesn't just determine where waste *goes* but also shapes our ability to attend to waste in the everyday and in theory. But what if waste didn't "go" away, if it weren't picked up off your streets, or if the street was where you lived? What if your home was described by scholars as "the gutted infrastructures of segregated cityscapes," in which the effects of imperialism reside in the "microecologies of [your] matter[ing] and [your] mind"?[47] What if waste was always becoming other things, including the logics that justify your removal, turning you into matter to be cleared and disappeared? Is waste, then, an empirical object or a condition of history? A description of matter or evidence of violent relations? The facts of production or a fiction of capital? "Away" is a violent spatiality that naturalizes, if not *reveres*, property above all else. It is a fantasy of environments stabilized—dare I say naturalized—by the inheritance of white ownership and the infrastructures that serve it (including the biocentric genders of capitalism). If we desire to forget waste, so too do we desire to forget the histories of enslavement and colonization's eco-social relations that have long buried their environmental risks "elsewhere," an elsewhere with which the material present is entwined.

ELSEWHERE IS EVERYWHERE
(TO A PROPERTIED WAY OF LIFE)

Throughout this book, I use the term *elsewhere* to articulate three interrelated themes: (1) how waste infrastructures displace and concentrate toxicity to an underthought elsewhere that is not only material but, mundanely, Black; (2) what waste tells us about property and, specifically, what I call a propertied way of life; and (3) the intersections of the emergent fields of Black geographies and Black ecologies that point us to the everywhere nature of "elsewhere." I see these three themes as necessary for thinking about the relationship between Black life and waste, if not the environment more broadly. The genealogy of scholars upon which I draw across these themes has taught me how to hold together things that are often in disciplinary tension with one another. So while the connections I elucidate in the following pages are central to how this book moves from one chapter to the next and to how I situate myself ethnographically, there are also things (namely, waste) that become more and more opaque as the book progresses. I see these moments of opacity as a critical intervention into ethnographic modes of attention and a way to insist that *seeing* is a struggle. I hope these moments remind all of us that opacity is also a political relation, a way of knowing violence, and a critique of Western knowledge production.[48] A critique of Western knowledge production that comes from somewhere else.

Infrastructures discriminate in the things they circulate, where they concentrate their social "goods," and, crucially, for *whom* they act. Against the anthropological framing of infrastructure as something that becomes thinkable or visible only upon breakdown, Robert Bullard's watershed *Dumping in Dixie* exposes how infrastructure's racist politics are not invisible. Breakdown is the condition of a racist capitalist state that affirms white lives while Black and Brown bodies are made to die.[49] By bringing scholarship on racial capitalism to bear on infrastructure, I start from the position that infrastructures stratify, distributing space, people, and services unevenly. They do not function equally in all places at once, nor are they meant to.[50] Urban spatial order is enacted through sanitation infrastructures (toilets, water treatment plants, sewers, and pipes), and ideologies of cleanliness and vitality are circulated and inscribed though practice.[51]

However, infrastructures do not just produce aesthetic and social orders; they also rationalize them into space.[52] Infrastructures materialize state power to coordinate and govern the spatial politics of everyday life.[53]

Joshua Reno's ethnographic work on waste labor in the landfills of Michigan argues that waste infrastructures "do not eliminate environmental health risks entirely, but *concentrate them elsewhere.*"[54] While infrastructures that "work well" recede into the background of white middle-class life, this is always contingent on the way they condition—that is to say, where and how far away they condemn the distant or unthinkable "other" to slow death.[55] Thus, while the research for this project began with questions about infrastructure, the book became an ethnographic and theoretical rumination on environmental risk as a mundane part of Black living. In this sense, the elsewhere is about marking that place and time to which risk is consigned, a way to insist that somewhere *over there* is in fact infrastructurally coterminous with living and theorizing *here.*

This brings me to the second thematic the elsewhere marks. If current paradigms for managing waste concentrate risk *elsewhere*, then what are the corollary spatial arrangements protecting the here and now from waste's accumulations? I argue that the here and now is categorically *property*, specifically white property and whiteness *as* property. Putting distance between waste and property is a critical part of settler-colonial, anti-Black land politics.[56] Moreover, to invoke Bullard's *Dumping in Dixie* once more, the structure of policies, practices, and directives "that differentially affects or disadvantages . . . individuals, groups, or communities based on race" is constitutive of racial capitalism's geographies and the material realties of colonialism.[57] In other words, it is *the possibility of ownership* that creates the conditions of dispossession, what Malcom Ferdinand has called "a colonial habitation."[58] Otherwise illegible to colonial forms of inhabiting, the environmentalisms of the colonized, dispossessed, and poor "other" are recast as "criminal," a problem of backward stewardship "over there." This colonial habitation, or what I refer to as a propertied way of life, disavows "the material, economic, and political connection to the 'here,'" and *here* is the time, place, and racial relations of property.[59]

The inherited "material privileges attendant to being white" define what Cheryl Harris describes as the secured expectations and status that make whiteness a form of property.[60] The recursive logic of property rights retroactively secures an ontological status to whiteness that secures the settler as a specific *genre of being* and defines his humanness by his right to first possession.[61] This ontological dominion is material, actively produced through the violence of conquest. I emphasize "material" here to center the way that social domination is always an environmental project and its violences an imposition on our complex ecological relations. As Frank Wilderson reminds us, *clearing* (to which I would add *cleaning*)—in terms of clearing the way for a propertied way of life—is a violent verb, laboring across Indigenous lands, bodies, and more-than-human beings in order to justify colonial improvements to "wasted" lands.[62] Critical Indigenous studies scholars teach us that declarations of "wastelands" are part of a colonial architecture that requires the genocidal removal of the native.[63] By marking Indigenous stewardship as "unproductive" and a waste, colonial regimes lay moral claim to ownership over land, non-human beings, matter, and all manner of peoples. As Indigenous stewardship, cosmologies, and ways of being with the earth were actively obliterated (at least according to the maps that ontologized settler sovereignty), the colonial appropriation of land became so thoroughly embedded in the settler-planter common sense that whiteness appeared to be a natural system that cleans, orders, and produces value(s).[64] However, property (as/and whiteness) is a force that requires waste to propel itself into being.[65]

While whiteness inherits the capacity to possess, whiteness as property becomes a *way of being*, marked by the "romantic marks of sentience [and a] feeling of intentionality."[66] Blackness, on the other hand, as Katherine McKittrick has argued, inherits dispossession and, along with it, the discourses of "dirt" and the unruly agency of unthinking matter.[67] Caught within the colonial gaze, contaminating nonwhite "others" are positioned as like objects or animals, justifiably slated for exploitation, manipulation, or early extinction.[68] Other genres of being within a racialized world become subdivisions of humanity—sub, less than, not human—presumed closer to inert matter, which must be enrolled in a recursive white regime of property rights to be made productive.[69] Property's conditions of avowal and disavowal are critical to understanding how waste *conditions*. Thus, waste management is not simply the infrastructural project to manage surplus matter; managing waste is an epistemological and ontological concern with the violent conditions of ownership.[70] Waste accumulations *become* toxic because managing waste, or rather our "wasting relationships," is part of the violent requirements for transformation that sustain the political, economic, and ontological conditions of white supremacy's propertied regimes of being. In other words, moving waste out of sight and out of mind sustains a settler-colonial fantasy: that whiteness and property are ecologically sound.

Thinking from somewhere else requires analytic tools that see facts, particularly the obviousness of space, property, and waste, as suspect categories of analysis produced by the circulation and exploitation of risk. However, under settler colonialism, as Tuck and Yang note in their theorization of the settler-native-slave triad, the racial formation of the native and the slave are distinct in their forms of dis/possession. The ways that anti-Blackness and the disavowal of Indigenous sovereignty require one another to transform land are strategically different, ensuring that their political incommensurability is a usable impasse for capitalism. Building off what Patrick Wolfe calls "the logic of elimination,"[71] I understand the terraforming of Indigenous lands with enslaved labor to be an ongoing process whereby anti-Blackness remains a useful way to declare land "wasted." The way that slavery codified anti-Blackness as, in part, a racial regime of expropriated labor enabled the expropriation of Indigenous land. Settler colonialism did not merely designate Indigenous stewardship as *wasting land*; rather, it enrolled slave labor in its *subsequent material expropriation by the planter class*. The political economy of plantation slavery under settler colonialism continues to shape how Blackness and waste are made proximate. Shifting our orientation to waste as a condition that links property to the elsewhere pulls into focus colonial and Black geographies, ongoing forms of genocide and slow death, and the necropolitical regimes of mapping and exploiting risk that make race and waste material bedfellows.

The third evocation of the elsewhere is the *everywhere* of Black ecological and geographic thought. Black ecologies names, in my estimation, the abundance of practice, theory, ways of being, and forms of critique in Black nature-social relations (or, my preference, material-social relations). Naming this abundance is itself an injunction to how the transatlantic kidnapping of African peoples then "planted" around the Americas and Caribbean cast Blackness *outside* nature-social relations—and outside Africa. By this I mean that the slave was used as a tool of ecological destruction to carve out colonial property regimes through which colonists could extract and then commodify land. But as long as enslaved people were used as a tool, enslaved people also resisted. Maroons sought to make life on their own terms.[72] And while some histories of marronage in the Caribbean are romanticized, it is important to keep in mind that life *outside* one "that is value constitutive" for the colony often requires experiment and improvisation, failure and revision.[73] Through the experiments that failed and succeeded (to mount revolution, to stay hidden, to live independent of colonial racial capitalism's monoculture), marronage is an exemplar of alternative

nature-social relations, other ways of being, and the alternative land relations that racial capitalism's ongoing colonial relations try to narrate as a
threat by calling them unruly. And they *are* unruly, as long as the rule of law
is private property.

The Elsewhere Is Black adds to the repertoire of Black geographic and eco-
logic knowledge that takes seriously how risk adheres in the matter(s) and
mattering of Black life. It takes seriously how the metaphorization of waste
requires a reckoning: both as an accurate metaphor for capital's behavior
and as an elision that sometimes distances Black scholarship from how
discards have long been part of the matter with which Black life is made.
To that end, I deploy Black feminist thought to guide our way through
unruly relations elsewhere, tracing objects as well as forms of objectification. By this I mean that within the domain of Black feminist thought,
objectification is simultaneously anti-Black violence's racial objectification and the sexual objectification and commoditization of racialized bodies subject to patriarchy's dominion.[74] Within the intersecting violences
of objectification, Black women are subject to a number of incongruent
contradictions that make them hyper-available and invisible, outside the
category "woman" and hypersexual, the "unthinkable" spatial subject, and
contagious in their procreation, to name just a few.[75] The architectures of
power that play out on Black women's bodies are deeply revealing of how
"wasting relationships" are also shaped by hetero(cis)sexist relations.[76]
Black feminist thought reveals how power wastes and how Black women's
inability (and refusal) to conform to white gendered expectations make
Black genders a site for material punishment.

Our feminist forebears have long theorized that colonization's gender regimes deem people of color gender nonconforming and sexually
deviant.[77] In the United States, "the master-slave relation constructed
a masculine power hierarchy" in which masters were "*the* representative
of hegemonic masculinity."[78] While post-emancipation racial violence,
such as lynching and castration, deliberately targeted Black men to reinforce white superiority, white masculinity's monopoly on violence shapes
the production of Black genders.[79] The settler-planter patriarchal order
turned Black genders—insofar as gender is part of what marks one's *full*
humanity—into an ontological impossibility. That is, enslavement required
the evacuation of the slave's humanity in order to produce manipulatable
Black(ened) "flesh" as a fungible object of exchange.[80] This "ungendering,"
as scholars citing Hortense Spillers have explored, not only subtended the
possibility of property in the form of humans but also turned Black people

into objects of exchange—objects upon which whiteness accrues inherited *material privileges* and disavows its waste.[81]

The punishing reality of "flesh 'ungendered'" is both critical to what C. Riley Snorton has called the fugitive "transcapability" housed within Black fungibility and the material anchor that whiteness and/as property requires to stabilize one of its central material fictions: that there are two and only two genders.[82] "A secret of cisgender," Marquis Bey writes, "is that it is not only about gender."[83] By this, I take Bey to affirm what Snorton argues: "Gender is a racial arrangement of the transubstantiation of things."[84] But these *things* that transubstantiate are the racialized matters of Black life and the material with which Black life is lived—whether that be on the run or in the hold. Important to my materialist approach is an understanding that cisgender, then, is not only a racial alignment but a racial and material arrangement of a colonial habitation—that is, a propertied way of life.

Thus, Black feminism—which in my curation requires Black trans theory and a queer of color critique—better traces the knowledge produced within confinement as well as the joy, aesthetics, pleasures, and genres of gender produced therein. This requires a queer embrace, as Cathy Cohen argues, of the non-normative Black lives that queer theory's critique of heterosexuality often ignores. For Cohen, "queer" should include "all those who stand on the outside of the dominant construction of norm[alized] state-sanctioned white middle- and upper-class heterosexuality."[85] In Cohen's analysis, this includes the deviance of single-Black motherhood and the fugitive sex workers and scavengers around whom this book's offerings pivot.

Elsewhere charts a Black feminist path through the "liberatory aspects of deviance" to espouse a non-respectable and queer-material ecological ethics of improvisation.[86] This ecological ethics sometimes evinces itself through the fugitive plans of the scavenger's hustle and sometimes through the carefully coordinated flight of sex workers through violent spatialities and white masculinity's monopoly on violence that continuously discards them.[87] I see these improvisations in critical conversation with practices of sabotage. For example, Sarah Haley rewrites Black women's sabotage in the postbellum south as a critical refutation of punishment's racial meanings. Black women's movements, joys, and right to rebellion violated the sanctity of whiteness: "Fugitivity was immanent, freedom ingrained in their interior lives even as the external world indicated they were trapped."[88] Discussing the planned, though thwarted, prison break of three Black girls incarcerated for destroying flowers along the paths to white houses, Haley writes, "The quotidian, deviant and gendered fugitive practice of floral theft and redistribution, the inspired collective imaginary" that inhered in rebellious friendship, fostered the "capacity [for] sabotage."[89] Haley's writing on sabotage, not as "success or triumph against systemic violence and dispossession" but as living by disordering, inspires how I see the ecological ethics that inhere in dispossessed Black life.[90]

Unruly places are abundant with unruly object relations challenging a propertied way of life. The abundance to which I speak is twofold. It is the capacity to generate life-giving critique and the capacity to use the (mis)names of matter to write new stories. In the case of waste, those lives discarded by racial capitalism make new forms of living that do not just replicate life as a value-constitutive activity for racial capitalism or the colony; rather, they alter *theory*. Centering altered lives produces alternative readings, including the alternative ways of reading sand, sexuality, and coloniality in the Caribbean in the work of Vanessa Agard-Jones; of reading US Southern environments through work songs in J. T. Roane's meditations; and of reading sorrow songs in the work of Willie Jamaal Wright.[91] This abundance is part of what Nik Heynen and Megan Ybarra call abolition ecologies.[92] The surplus matter of racial capitalism (waste in its many forms) and the surplusification of life are not things that can ever be made pristine.

Centering low-income, poor, and houseless material practices as *material knowledge* requires thinking with the unruly abundance of matter produced by the "wasting relationships" that discard. After all, it is crucially *with* and *into* unruly environments that enslaved peoples forged their escape.[93] The alternative material-social relations that enslaved people built with their own human flesh and the flesh of the earth challenged the plantation's disposal of Indigenous homelands. The "unruly," "deviant," "dangerous," "waste-filled," "bad" environments where white men (and women) dare not go just might be the ecological conditions facilitating forms of resistance that provide shelter from the hazards of a heteronormative, gender-austere, racially violent, propertied way of life.

STORY IS THE PRACTICE OF BLACK LIFE

Inspired by the people whose words move us from page to page (from theorists like Fanon and McKittrick to my interlocutors Betty and Sal), I, too, try to experiment with living Blackly on the page. I improvise where theory fails, I reach for other objects where waste overdetermines, and I weave speculative histories where "proof" does not exist. Deeply informed by Black feminist thought, I revise—which does not mean stably assert—to perform the kind of still-movingness of scenes of Black life.[94] I take to heart McKittrick's assertion that alongside the archive of slavery, the way scholars often bring Black life to the fore is to render the violence anew. As such, "The documents and ledgers and logs that narrate the brutalities of this history give birth to new world blackness as they evacuate life from blackness."[95] It happens in ethnography, too. Thus, I want to be clear from the outset: Ruth Wilson Gilmore's definition of racism as a condition of vulnerability to premature death remains true, and waste, in this book, is part of that condition.[96] But this book is also concerned with how people *live*, how people manage both the matter and the *meaning* of being discarded.

The anti-Black logic of capital's dispossessing migrations means that ethnographically, I had to learn to write about people I met amid the constant loss of them. Learning through constant loss is a fundamental challenge to the project of ethnography. The viability of the ethnographic project hinges on a racial-spatial imaginary: presumptive access to "the native," to which the ethnographer can continuously return. But dispossessed Black people living precariously on and off the street are always subject to violent removal. In other words, my interlocutors were always moving. This posed a methodological challenge to feminist ethnographic practices that demand reciprocity and ongoing intimacy beyond the "end" of research or writing. Instead, my feminist training pulled me to think about the relations of power that inform knowledge production and the epistemic violence we as scholars commit in our role as landlords of knowledge. As landlords, Joy James argues, we authorize the ideas that circulate for and alongside capital.[97] This indictment is a challenge not only to academia writ large but also to our writing, our commitments, and to what and who "research" is for.

So many of the people who inhabit these pages disappear (and are disappeared), so much so that when I began to write this book, I thought it was fundamentally a book about loss. But in writing, imagining what it might be like to return to conversations, to ask people to read what I'd written, or to learn where our voices are more dissonant than they seem on the page, I've realized that this book is an homage to the lessons I've learned from Black studies as the practice of Black life, a celebration of experimentation, improvisation, and speculation as tools for surviving anti-Blackness. To square my feminist training with the realities of Black life, I had to find a way to honor how dispossession constantly produces people and theory on the move, including to prison.[98]

Johannes Fabian famously wrote, "Coevalness is anthropology's problem with Time."[99] But ethnography is also guilty of a kind of presentism that assumes that "informants" never change, have no history, and worse, do not have changing conceptions of their own histories and selves. Changing self-knowledge is, I believe, fundamentally inaccessible to the project of ethnography but is also necessary to confront in ethnographic writing. My approach here is not a whole answer but a strategy (one that I hope continues to change) for thinking while writing about dispossession. Drawing on the political meditations of Terrion Williamson and reflexive theorizing of Saidiya Hartman, I actively write myself in.[100] In lieu of returning to people during research, I return over and over again to them on the page. Across the book, I restage scenes, returning to wonder what other ways and what other things were known—*wandering*, as Sarah Cervenak might say, toward "an undisclosed terrain" of ecological ethics and desires.[101] Without access to those changing reflections, I instead return to a multitude of conversations. I revisit interactions over the course of the book that have stuck with me, and my thinking about them evolves (and will continue to evolve, I'm sure, long past the publication of this book).

LIVING BY DISORDERING

The interlocutors in this book are criminals. They break the law, and they do so with pleasure.[102] They steal things and swipe pills; they destroy property. Some announce affinity with trash and pestilence, some speculate about toxicities that can't be proven and the histories we can no longer see. They make fugitive plans, reclaim trash as matter that matters. They fabulate and experiment, improvising with objects and claims to them. They disappear, and they are disappeared. They are disreputable, and to some, unrepresentable, self-proclaimed trashy women. They are pedagogues of an anti-respectable environmental politics where unruliness is a different horizon of relation and ecologies are always marked by the violence of capital. Never pure or clean, they teach us other things about the environment, other ways to be *against purity* and to challenge the assumption that property is ecologically sound.[103]

As I previously noted, this book disorders. It follows fugitive histories, and waste becomes fugitive matter. Because this book foregrounds the knowledge produced by living while Black and on the move, sometimes scholarly approaches to waste sit incongruously with how people describe their life and relationships to discards. For this reason, between each chapter you will find short meditations on terms that often characterize (disciplinary) approaches to waste. These meditations deploy different tactics for speculating about how each term is connected to Black life. In other words, these interstitials put Black and discard studies in conversation with one another. "Flow" looks to the importance of a waste contract signed in the 1990s that forever yoked New York City to Virginia infrastructurally. "Infrastructure" thinks about how the plantation haunts the scale at which we manage waste. "Surplus" shifts our attention from the politicization of specific objects to the way waste is a different kind of surplus critical to racial capitalism, and "Disposal" thinks about how Blackness becomes necessary to dispose of things in the first place. In each vignette, the 1990s plays an important role, making these meditations a place to wonder what actors and mandates are involved in turning plantations into dumping grounds. But in the final vignette, "Junk," waste becomes a place to wonder what else happened in the 1990s, not to make an argument for better environmental policy but to notice what existing environmental regulations occlude and what Black people make with waste's constraints. These meditations can be read together and continuously with one another or as interstices, points of connection, and moments of disjuncture that emerge when the environment is storied from here, there, and elsewhere.

Chapter 1, "Toxic Capture," thinks through how waste becomes a form of ecological punishment, and it does so in part by reframing the common assertion that "Black people don't talk about the environment." Instead, this chapter argues that anti-Blackness is an environmental strategy, showing how stories about surviving racism, criminalization, and dispossession are in fact stories about the economies of waste shaping the environment. The criminalization of poverty, which positions the poor as at fault for their own ill health, obfuscates how toxic conditions are produced by waste-management's protection of white propertied life. Black people are presumed to be inherently hazardous (socially and materially), providing justification for the toxification of the places where Black people live. Thus, living proximate to the municipal, regional, and private facilities that infrastructurally coordinate waste's movements, means being criminalized as a hazard to property value. If white supremacy's values determine how land is "used," it does so by turning Black people's bodies into a material threshold for environmental risk. By centering those people who are (unknowingly) inundated by the long-distance management of New York City's waste in the Virginia Tidewater, waste, in the present, is revealed as a geography of toxic capture.

"Becoming Fill" (chapter 2) locates toxic capture as an outcome of the racial and ecological injury of the plantation. Tracing how wageless sex workers in Virginia tell stories about their own surplusification, this chapter uses their descriptions of land and matter to excavate chattel slavery's and settler colonialism's ongoing destruction of Black flesh and the flesh of the biosphere. By emphasizing that "wagelessness" is a relationship to (native) land, this chapter shows how Black people's bodies (not just their labor) are rendered "unecologic" and excess matter central to capitalist development—the recurring terraforming projects that fuel settler colonialism.

Chapter 3, "Revisions from Elsewhere," flips the script. If waste is used to Blacken environments, chapter 3 asks us to see the landscapes of waste as a possible condition for radical Black becoming. This chapter takes us around town and around time, not only to those places where waste marks fugitive movement but to those times where trash and identifying with it become ecological protest. Raising questions about what lessons we might take from trash, waste becomes part of the fugitive, unruly matter of Black politics and Black life. "Revisions from Elsewhere" presents a challenge for thinking about Black futurity and the respectability of environmental politics. Here I trace a dirty Black feminism to open up a place of intersectional politics, where those discarded by racial capitalism's wasting relations have something to teach us about living with trash.[104] They offer a radical imperative to a future already shaped by our past and present waste.

Chapter 4, "Black Refractions," transports us to New York City and the houseless Black men scavenging Bed-Stuy in the midst of gentrification. Amid the legacies of spatial violence that continuously avow and disavow Black history, discarded objects take on new importance. Both as ephemera of the impossibility of the Black "domestic" and objects of potential value, reading among the litter of Black life asks us to tend to the relationship between Black genders and place.

The conclusion, "Fictions of Fabulous/Fabulative Ethnography," is a playful meditation on the tensions the preceding chapters raise about story, time, Blackness, and ecology. These final experiments don't summarize where we've been but point toward a place our ecological politics could go. In other words, what do people do with the elsewhere? How does story become Black life and Black life become study? What is a Black eco-grammar but a grammar of living (improvising) *against property*?

Neither progressive—at least, not a progress narrative that moves from savagery to civilization—nor pure, in the sense that categories of matter are not always what they seem, the story I tell of myself eating books is not a clean one. And neither is this book. *The Elsewhere Is Black* is not a story that promotes cleanup or a story that denies that poverty is dirty. It is a discordant story, in which progressive environmental paradigms refract the long duration of anti-Black settler-colonial property regimes and the genders they seek to impose, while also considering how identification with waste and dispossessed people's experiments with its meaning and matter are sometimes a declaration of anti-liberal personhood, sometimes a way to notice living history as well as underthought sites of political becoming. Like the story I tell of myself through family here, this book is a story about *dirtiness* and *trashiness* as knowledge, Blackness as an alternative site from which to see the "environment," and Black being as ecologically complex. It revels in the complexity of the toxic, tacking back and forth between the violence of untraceable sources and the ways people story it. *The Elsewhere Is Black* also tracks the femme aesthetics of "unnatural," trashy women to forms of ecological punishment. Trashy women's environmental politics refract a different relationship between property and gender, one that hinges on a "criminal" irreverence for property. Throughout these Blackened places, time and again people theorize how environmental racism *shapes* but does not *determine* Black living. Refusing to reconcile the toxicity endemic to racial capitalism, this book is an ethnographic exploration of ecological modes (real and speculative) forged elsewhere. And that elsewhere is Black.

FLOW

On July 7, 1997, the residential trash of Bronx County, New York City, began traveling to Virginia. This was not, as Mayor Rudolph Giuliani had framed it for concerned activists, an environmental win. New York City's infamous Fresh Kills landfill had been politically noxious, and diverting trash from it was a way to curry favor with the city's whitest borough, Staten Island.[1] In a press release, Giuliani's administration proudly announced: "Under the three-year, $86 million waste removal contract awarded to Waste Management Inc. (WMI) earlier this year, WMI will transport 1,700 tons of residential trash from the Bronx each day. . . . [The mayor noted,] 'In 2001, images of waste from all five boroughs being dumped on Staten Island will no longer be a daily headache. They will be memories.'"[2] But as environmentalists and city council members pointed out, Giuliani's borough-based waste plan had merely displaced the city's waste problem.[3] Signing a contract with a private waste hauler to transport the city's trash merely redirected an acute reality. After all, the city's waste had to go somewhere.

For fifty-three years, the city's waste had gone to Fresh Kills, known colloquially as the Staten Island Dump. And the dump hosted residential waste from all five of New York City's boroughs. Initially opened in 1948, Fresh Kills was supposed to be a temporary solution to the closing of the Rikers Island landfill. At the behest of Robert Moses, the infamously racist city planner and chair of the Triborough Bridge and Tunnel Authority, two thousand acres of "worthless" wetland in Staten Island was turned into a holding place for waste.[4] For Moses, this temporary dump was central to his industrial plans, a way to clear Rikers Island for development.[5] However, Rikers Island didn't become a site of *industry* exactly; rather, it became New York City's most notorious jail. And Fresh Kills was not temporary; it lasted for half a century. By the time Giuliani came on the scene, Fresh Kills was the sole operative landfill for all the city's residential waste, and it had become Staten Island's rallying cry for secession. The two-term mayor was staunchly revanchist; committed to bringing back an imagined (white) glory to a "white middle class that sees the city as its birthright."[6] A trash plan was critical. It didn't matter where the trash would go as long as it was gone.

But for Virginia Senator William T. Bolling, one of the five state senators to sign a letter contesting Giuliani's trash plan, it *did* matter: "Virginia is clearly the target and intended recipient for the vast majority of New York's trash."[7] In December 1997, alongside the initial 1,700 tons of trash from Bronx County, another 1,970 tons of Brooklyn's trash was added to the daily delivery, and by 2001, the number was intended to rise to 3,900 tons of solid waste per day. On top of that, the plan estimated adding another 2,600 tons from Manhattan and an unspecified amount from Queens.[8] By 2012, waste exporting had become the state's primary long-term waste management strategy, with estimates that the city exported thirteen thousand tons of waste to Virginia every day.[9] When the Fresh Kills landfill officially closed in 2001, the city's "memories" had become an elsewhere reality—an elsewhere in which Tidewater, Virginia, was entangled.

DIRTY IS THE SOUTH

This flow of waste—from New York City to Virginia—was made possible by a racist political economy as much as it was by Giuliani's WMI contract itself. Where waste goes is a measure of the racial metrics of property value and a fiction of capitalism's environmental control. The management of waste indexes who and where are valuable to the prevailing political imaginary. Inside Giuliani's revanchist policies of the 1990s, the *who* and *where* that mattered is particularly clear. In the "vacuum created by the dismantling of liberal urban policy since the 1980s" and white flight, Neil Smith argues that the liberal assumption that government should aim to ensure some kind of minimum quality of daily life was turned into "a vendetta against the most oppressed—workers and 'welfare mothers,' immigrants and gays, people of color and homeless people, squatters," and squeegee men.[10]

This attack on the mere *presence* of the poor (and poor people of color in particular) was a concerted effort to identify who had "stolen" the city from the white middle class.[11] Of course, it was not the poor but the dismantling of urban policy that began in the 1980s, which produced the capital *flight* and fiscal crises that followed. With the poor (mis)identified as a hazard to the city, the police became the arbiters of "cleanup," charged with "sanitizing the landscape." Giuliani believed that criminalizing poverty "would reverse the urban decline, opening up the possibility of a new city."[12] Under the guise of a moral imperative, sanitation became a metaphor for policing and "broken windows" a construction of "criminal environments."[13] Poor neighborhoods bore the sign of "dirt and disorder," the progenitor of quality-of-life "decline." And this metaphor was materially reinforced by the trash plan; after all, sending waste south did improve the quality of life for those who owned property, especially for the white middle class in Staten Island. For property owners, the threats to their quality of life were disappearing, and Giuliani seemed to be right: "The symptoms *were* the cause."[14] As Giuliani's infamous broken-windows, stop-and-frisk, and anti-graffiti policies violently targeted poor people of color as "decay," New York City's *actual* trash was quietly finding its way down to Virginia.

With a clear ideological "cause" for the city's decline proffered and hammered home in city policy and popular culture alike—meaning the presence of poor people of color in neglected neighborhoods—restructuring city services provided an opportunity for Giuliani to outsource an infrastructural problem, trash, to a private company: Waste Management, Inc. For a private waste hauler, New York City's waste was a game changer. The city's constant stream of trash was a corporate dream.

As Giuliani whispered promises of never-ending contracts to WMI and rebuffed city council and grassroots activists' concerns about displacing the environmental risk of waste somewhere else, Giuliani was making a calculation between the cultural "backwardness" of the south and the environmental role it ought to play in New York City's progressive future.[15] In fact, in what can be best described as a showdown between the city's then mayor and Virginia Governor Jim Gilmore, Giuliani made it abundantly clear that this waste route was not temporary. Sidestepping questions of the "environment," Giuliani implied that in exchange for the culture of New York City, Southerners should welcome the city's trash: "We don't have room here to handle the garbage produced not just by New Yorkers, but by 3 million more people that come here, that utilize the place everyday. . . . People in Virginia like to utilize New York because we're a *cultural* center, because we're a business center."[16]

Despite Governor Gilmore's push for an out-of-state waste cap, interstate waste export is protected by the Constitution.[17] And, it should be noted, Jim Gilmore's campaign budgets, like that of many Virginian politicians in the

1990s, revealed significant gifts and donations from the waste industry.[18] Ultimately, Virginia's attempt to curtail the amount of trash imported from out of state failed. On June 4, 2001, the US court of appeals upheld the Virginia district court's February 2, 2000, ruling that statutory provisions enacted by Virginia to cap out-of-state waste imports violated the Constitution's Commerce Clause: States are not allowed to ban waste shipments from other states.[19] The Commerce Clause, which protects the flow of goods across state lines, is often a vehicle for corporate influence over regulation, paving and protecting its own corridors of flow and profit, environmental consequences be damned.

While Giuliani felt that he had secured a solution, it might be more accurate to say that wmi had cornered the market. Within a few years, wmi would sign a total of three contracts with the city to export its trash to the company's mega landfills in Virginia, as well as sign a contract with McAllister Towing and Transportation Company, a water transport hauling company; purchase $10 million worth of airtight containers (for shipping waste); and invest $15 million in the expansion of the James River trash facility in Virginia. By 1993, wmi had constructed seven mega landfills in the Tidewater region, ushering in unprecedented tons of waste.

Waste in Virginia is no metaphor; Virginia is the second largest importer of waste in the nation. There are now eight mega landfills in Virginia, with the Atlantic Waste landfill in Waverly, Virginia, being one of the largest in the country. Distinct from the municipal landfills operated by local governments or regional cooperatives, the waste flows of mega landfills direct local and state regulations in their favor for their own profit.[20]

Making waste flow birthed a multibillion-dollar industry, and the direction of waste's flow tells us something about how settler colonial racial capitalism manages the environment.[21] While there are landfills all over the United States and New York City does send waste north (to the Finger Lakes region in upstate New York), the concentration of landfills in the South is undeniable. Is it a coincidence that these flows seem to map the reverse course of the Great Migration? Or is there a more sinister reality: that where Black people are concentrated, so too is pollution. Perhaps in addition to what the Great Migration maps—the environmental formations that facilitated hiding, escape, and fugitive movement—it also maps a relationship to material risk, including past and present waste flows.[22] And perhaps this flow of waste is an environmental form of punishment wrought by that "peculiar institution" that set off the Great Migration. Perhaps this flow is not a "coincidence" but a geography of capture, an afterlife of the ecological relations of slavery and the resilience of what Clyde Woods calls "plantation relations."[23]

The flows of risk and forms of capture that shape Black life have always demanded that Black people continuously engineer new material relations, new ways of (at)tending, and new ways of thinking against their own capture. "The enclosure of southern places, which set off the dislocations of the Great Migration," J. T. Roane writes, was also accompanied by "the concomitant reterritorialization of Black ecological practices and knowledges."[24] Post-emancipation, Black people brought alternative forms of material praxis that facilitated their survival on and off the plantation to urban contexts. But as they plotted in new urban places, sowing the seeds of alternative use values, Black material practices were "reduce[d] to the folk or inscribed as dangerous."[25] As the anti-Black violence that conditions the present suggests, these ever-evolving life sustaining practices are criminalized, or as was popular in the 1990s, charged with the degradation of the white city. Roane encourages us to hold onto material practices of survival as a well of Black eco-theory, generative of alternative ethics, the ethics of a Black commons. Out of the condition of enslavement Black people brought with them a different orientation toward matter as a necessity, allowing them to intervene in and reconceptualize new places as home.

But the continuously evolving spatial regime of racial containment seems to "burde[n] down [the same places] with a freight of symbolism" that is as heavy as the tons of waste that seem to follow.[26] The corridors carved by Black migrants' insurgent desires are the same as those now filled with "an average load of 4,000 tons [of trash] each weekday."[27] The routes once imagined as paths toward freedom are now arteries through which toxic tethers tie people, land, and aquatic life to a Black condition—the condition of being "elsewhere."[28]

. . . .

The way waste flows—where and toward whom—is not an innocent project.[29] And the operative assumptions about race, land, place, and value materialize a violent truth. Whiteness "operates possessively," and "as a possession it enhances one's life chances as it is configured through the logic of capital."[30] And Blackness is a structuring force that can always be evacuated of meaning, particularly ecologically, allowing whiteness to operate possessively to enhance life chances. The places that are linked by flows of waste are tightly woven together by geographies of Black capture: the historical capture of Black labor, from which the plantation's logic continues to shape *place*; the racial formations that undergird "land use," from hazard siting to real estate projects; and the ramifying health effects of living with the risks of waste infrastructure, not to mention that jails, prisons, and detention centers overly represented by people of color are built on top of Superfund sites and landfills, or are sanitation catastrophes that double the toxicity of prisons themselves. Flows of waste articulate something that the plantation anticipated: that Blackness is infrastructurally productive of whiteness's grammar of property and possession while simultaneously tasked with absorbing settler colonial racial capitalism's toxic way of life.

1 Toxic Capture

The current environmental protection paradigm has insti-
tutionalized unequal enforcement; traded human health for
profit; placed the burden of proof on the "victims" rather than
on the polluting industry; legitimated human exposure to
harmful substances; promoted "risky" technologies such as
incinerators; exploited the vulnerability of economically and
politically disenfranchised communities; subsidized ecological
destruction; created an industry around risk assessment; de-
layed cleanup actions; and failed to develop pollution prevention as the overarching and dominant strategy.
— BULLARD, *ENVIRONMENTAL JUSTICE IN THE 21ST CENTURY*

"It always comes," Darius lamented. We were sitting in a coffee shop in
Ghent, a newly gentrified neighborhood in Norfolk, Virginia. Darius
worked with incarcerated youth, some of whom were from Norfolk, all of
whom were Black and Brown. It was November 2015, and he had just finished a visit to the projects in St. Paul's, a neighborhood a little over a mile
away. Gentrification's many terms for dispossession—*urban development,
urban renewal, neighborhood rejuvenation, regeneration,* and *city beautification*—
"shape how my boys can or cannot live," Darius said. "This, where we are
right here, it used to be the hood. But the hood was cleaned, leveled. Blacks
were dispersed across the Tidewater. It's a history we all know, you know?
Slum clearance, block-busting, redlining . . ." Trailing off in the list of spatial violences he could name, Darius leaned back and sighed. Then, as if to
press his words into the white atmosphere of gazes that lingered a little too
long on our table, he sucked his teeth and said, "You know us n****s out here
just been tryna rest since we left the plantation!"

Uncomfortable, onlookers quieted to see what would happen next.
Would we upend the terms of white civility inspiring the police brutality Teona Williams writes about when she says, "Violence [is] a form of

environmental control"?[1] Would we be removed, just like Black residents in East Ghent were in the 1960s and 1970s, when the conflation of degraded buildings and degraded people inspired the land grab that turned Ghent into a home for the quiet coffee shop in which we now sit? Would someone decide that our Blackness was fundamentally at odds with their right to "peace and quiet"?[2] After all, Virginia has some of the most permissive gun laws in the nation (if you're white). "Shoot," Darius said, sucking his teeth. He quieted down, leaning back in his chair and sucking his teeth one final time. We both understood the terms of our existence in this (now) all-white neighborhood on the all-too-appropriately named street, Colonial Avenue.

Darius continued to describe Norfolk—"It's half white and half Black, just like all the major cities around here. The two sides live very differently"—and I thought of Fanon's description of the "two worlds" of colonialism: "The settlers' town is a strongly built town, all made of stone and steel. It is a brightly lit town; the streets are covered with asphalt, and the garbage cans swallow all the leavings, unseen, unknown and hardly thought about. . . . The settlers' town is a well-fed town, an easygoing town; its belly is always full of good things. The settlers' town is a town of white people, of foreigners."[3]

Just as Fanon describes the settlers' ever-expanding town—a town organized by the lines of property, where settlers' aesthetic practices (brightly lit, easygoing), object relations (garbage cans and bellies full of good things), and materiality (stone, steel, and asphalt) are enshrined as culture and law—Darius described the Hampton Roads as someplace set apart from the poverty of the Tidewater. "The Hampton Roads," he said, when I asked him to describe the difference,

is where the whites live in their pretty little homes and their clean yards, as if it's something different from the tidewuddah where the poor live. It's like they put the trash out but actually it just comes to us. Like it *is* us. Like in Suffolk, there's the landfill and Portsmouth there's that fucking "green" plant that white people love. You know? But it's also just everyday shit. Like in Ghent, you're in the fucking hood and there's shit on the streets, you know, trash, rats, whatever, tree roots busting through the sidewalk, houses falling down . . . and then you turn a block and it's like where the fuck did this plantation come from? It's like two worlds, two times, maybe like a whole other fucking dimension. It's like you just walk one block and BAM! There's the good life slapping you in the face. You best believe it's a slap too, they'll call the cops on your ass so fast.

The white place of the Hampton Roads, with garbage cans that "swallow all the leavings, unseen . . . and hardly thought," is enabled by the ecological relations of settler racial capitalism. The leavings of whiteness (as property) condition the elsewhere(s) of the Tidewater, the hood and the plantation, the "green" plant and the trash that always "comes to us." The disavowed use of Black bodies to place waste is a material punishment that uses the environment like the butt of a gun.[4]

More than environmental racism, an analytic that points to acute sites of disposal and the politics of siting hazards, Darius redirects our attention to the multidirectionality of ecological violence in the places of criminalized Black life:

> You know, it's not one specific thing, not one contaminant you can take out of the water. It's not one isolated incident that you rectify. *It always comes for us.* And it comes from everywhere. It's lead in our homes, in the water, in the food. It's garbage that the city never picks up. It's no heat in the winter, no food. It's insect and rodent infestations. It's eminent domain every time some asshole wants to "develop," by which I mean, *break* the things our communities persistently try to make. It's fucking capitalism; it is racism-induced poverty. Jim Crow never went anywhere here. It's a well-kept public secret, you'll see if you're here long enough.

Segregation is a form of colonial land use, and Blackness (because it's always displaceable, dispossessable, and disposable) is a threshold where toxicities compound.[5]

Darius's poignant phrasing, "it always comes," became a refrain through which to see the Blackening of environments—or rather, to see anti-Blackness as a land management strategy and waste as a racial condition of geographic condemnation. "The violence is everywhere, it's woven into the fabric of our lives, what we eat, where we live, the lack of care we get from the doc to the welfare office, just one barrier type thing to another. It just piles up." It accumulates. As we talked about the mundane ways that waste seems wrapped around or proximate to criminalized environments, I heard Darius, like Fanon, articulate what I now call "toxic capture": the ecological punishment that accompanies the geographies of poor Black life.

Waste is tangled up in Black life, not as metaphor but as material reality. And in the United States, where anti-Blackness is a land management strategy (Black codes, Jim Crow, redlining) enabling the ongoing theft of unceded native territory, disposal sites are themselves products of the colonial relations (political, economic, and geographic) that dispose. Though the distance from sites of disposal afforded by middle-class habitation protects some from the ideological and material effects of waste management's accumulative harms, waste infrastructure maps itself across the disavowed *use* of Black bodies to place hazardous accumulations. Moving away from thinking waste as primarily an object defined by its sites of disposal, I move with Darius and others to reveal waste as a material (environmental) condition of racialized condemnation, inventoried by criminalized Black life.[6]

To put it differently, proximity to waste is an environmental use of Black bodies. Sometimes this environmental use means their positioning as "excess" if not contagious matter slated for inexorable removal. For example, "slum clearance" means not merely the razing of neighborhoods but the displacement of Black communities to toxic elsewheres, including the toxic elsewhere of the prison. Sometimes this use is in the ceaseless pursuit of "cheap" land, where the presence of Black people recursively justifies the value of land as already worth less and thus available for seizure, from gentrification to locally unwanted land uses (LULUs). In the case of waste and its management, both private and public waste management rely on the violent conditions that make land available for waste's future emplacement. And this relies on the way that colonial racial capitalism expropriates Indigenous land to *degrade* and *pollute*. In Tidewater, Virginia, where landfills map onto former plantations, fungible Black bodies have long been enrolled in terraforming, producing, extracting, and toxifying environments. The conditions that make stolen land "cheap" are the conditions that cast Blackness outside nature and make *Black* and *trash*, as Charles Mills once said, redundant.[7]

Toxic capture is thus an ecological condition of land and the Black body that concentrates pollution to a *Black* elsewhere where risk can always accrue. As Max Liboiron argues, the threshold theory of pollution makes a colonial calculation: The capacity for *things*—people, land, and our more-than-human environments—to "safely" assimilate pollution "strips away the complexities of land" in order to produce a "resource for waste disposal."[8] This colonial science of "safe" assimilation relies on the thingification of "the other," its instrumentalization in service of white settler prosperity.[9] I want to draw our attention to how this threshold positions "the captive [Black] body as a vessel for . . . us[e]," what Saidiya Hartman and others refer to as the fungibility of Blackness inaugurated by transatlantic slavery.[10] Blackness, then, is environmentally useful precisely because there is no risk too great for Blackness to absorb. As "excess" to the environment, Blackness is made to absorb that which threatens white life on the one hand, and "pristine" nature on the other. Toxic capture thus expands our repertoire for understanding the expansive and inexhaustible use of Blackness beyond *labor*—for the pleasure of domination; to provide spatial, ecological, or economic "fixes"; and to *absorb pollution*.[11] Toxic capture reinforces the contemporary ecological threshold as continuous with the long history of suffocating Black life inaugurated by the slave ship's "hold."[12]

From the hold to the threshold, captivity is always by the settler-capitalist's design. Perhaps captivity is even infrastructural. As Simone Browne reminds us, it was on Jeremy Bentham's journey to conceiving the Panopticon in Krichëv, Russia, that eighteen Black female slaves were held captive "under the hatches."[13] By asking us to see the Panopticon—a schematic that sought to perfect surveillance through the architecture of incarceration—from "under the hatches," Browne, as I read it, forces us to rethink the relationship between efficient infrastructures (above) and the submerged captive body (below).[14] Underneath Bentham's feet were the wretched conditions of the Middle Passage, in which Africans chained together died by suffocation. The conditions of containment below served to *make* the slave an object of use, while those above provided the breathing room for Bentham to ensure that white men's authority was material, architectural, and infrastructural by design. With Browne's prompting, we might ask: Under the weight of infrastructural efficiency, who is submerged; whose life is held captive by the waste infrastructures that protect and construct white life?

Across the different places we encounter in this chapter and the people who story them are the experiences of what it's like to be cast outside of "nature" and instead held captive as the acceptable threshold where toxicity accrues.[15] Though it appears differently in people's stories of place, waste is always an axis that reveals punishment—the pathologizing and poisoning of place and the criminalization of behavior—to be an *ecological* relation. To hear waste as ecological punishment, I had to think alongside the contradiction that "Black people don't talk about the environment" would often be the first line in people's stories about the toxicity that penetrates Black life. This contradiction reveals "the mundane and persistent ways in which "death and perhaps extinction always already constitute existence for the 'fungible' object/being."[16] Black *being*, then, to quote Willie Jamaal Wright, is rarely treated as "the potential source of . . . new human–environment relations," making proximity to waste seem natural to Black poverty and casting poor Black people as outside "nature."[17] The threshold of toxicological concern not only is determined by the racial calculus of profit but reveals how capital's demands (mis)name poor Black people as a place to site "acceptable" risk. So, to reprise where Darius asked us to start, How do Black people theorize their lives when toxic capture is *always coming*?

ALWAYS AND NEVER THERE

"Wha'd he say? Play it again?" Mary yelled to me as I put the car in park. "Virginia" by Clipse, a hip-hop duo from the region, spilled out of my car window and into the motel parking lot.[18] With the car idling, I turned up the sound and skipped the track back so that she could hear Pusha T spit the lyric once more: "Virginia is for lovers, but trust there's hate here." As a self-described "mother, sister, hooker, and an all-around bad ass Black biiiiitch," for Mary, the verse was particularly poignant. "That's right! Trust there is *hate* here," she said, snapping her fingers as she nodded her head and pursed her lips affirmatively.

Mary's affirmation of Clipse's reference to hate makes explicit that the state's slogan, "Virginia is for Lovers," not only papers over the history of miscegenation and the vexed illusion to *Loving v. Virginia* but also romanticizes the present of Southern living as if racism were in its past.[19] "Don't get me wrong, I looooove the South, can't imagine living up north. But it's hard out here. Everyone's sick, in my family anyway. It's like a quiet hate spreading. I'm not quiet. I'll look a mother fucker right in they eyes and call out some fucking Klan bitch. Shit's racist as fuck. People in my family be dying, we don't know why. People in my family be getting cancer. Ain't no *love* for us."

Although Mary said she doesn't "really know much about environmentalism racism [*sic*] or whatever, don't think anyone gives a crap about that around here," her descriptions of her hometown (Suffolk) and Black life in the South suggest otherwise. In Mary's words, "If it's not the cops killing you, it's your food, your damn house, or that bad soil—ha!"

Mary and I had only a few conversations where it was just the two of us. Often when I talked to Mary it was alongside the other (working) girls at the motel: Jane, Betty, and others who prefer not to be named.[20] Betty was the first to tell me she noticed me, though I imagine it was hard not to notice me when I showed up at the motel. I was searching for evidence that being disposable is part of being Black in all the wrong places—in failed interviews with landfill engineers, at the waste treatment site, at the waste transfer station. I had been oversaturated by (white) vitalist arguments that the agency of "things" had been undertheorized, arguments that couldn't imagine that the Black experience of thingification might have something to offer.[21] The colonial heart of anthropology's whiteness can more easily imagine "a new materialism" in which things are lively and agentive than it can sit with the unruly agencies, material knowledge, and mattering of the social life of social death.[22]

The girls must've presumed I was mad. What Black woman goes looking for trash if they don't have to? Sometimes Betty would cock her head and say as much, "You all right?" like I was on the verge of (falling off? jumping off?) something. Jane would say, "Girl, WHAT?!"to my insistence that they had taught me more than site visits to waste transfer stations ever could. And Mary, too, would look at me the way my grandmother used to when she thought I was about to make a bad decision. But questioning my sanity—or perhaps it was that the girls identified a trashiness that lurked alongside my suspect madness—made it possible for the girls at the motel to invite, if not embrace, my presence. Thus, though it struck them as strange that I was "looking for trash," the girls were also adamant that I had already found it. For the girls, "trash" was what they were. And they would say it proudly: "ghetto trash," "hood rat," "trashy ass"!

Though the new materialists were turning everything into an object lesson, *being* Black is itself an object lesson of sorts, a punishing way to learn about how central the "racial" in *racial capitalism* is to land use, to capital, to property and propriety—not to mention to the (un)marked authority of whiteness in the composition of the *credible* narrator. The landfill engineers who had been excited to be interviewed and the local archivists who *on the phone* welcomed the opportunity to share what they knew hid behind doors, averted their eyes, refused an interview, or pretended to be someone else when I showed up in person. My Blackness always interrupted a new materialist methodology, requiring me to follow the object (waste) and interface with the imagined "experts" of waste's logistical coordination. Attempted observations at the waste transfer stations, phone calls to the municipal landfill, and persistent spontaneous arrivals at the places where waste is managed technocratically *while Black* further cast suspicion on my search for the thirteen thousand tons of waste that New York City exports south daily.

Searching for waste while Black and rebuffed at every turn by white experts whose certainty that *Black* and *trash* were redundant made a Black gaze *onto* trash impossible. It wasn't just the girls who thought I was mad; I did, too. And my madness was heightened by trying to find a place to live. The racist lending practices of a supposedly bygone era were well established in the Tidewater. My status as a researcher could never seem to mitigate my Blackness (much less my Black womanhood) and turn me into someone credible. Closed door after closed door is its own kind of spatial relation, and its own lesson into the uses of the (Black) object to produce the vitality of white life. Like Darius said, Jim Crow never went anywhere. And in 2015, on the verge of Donald Trump's first election, white nationalism's toxic swell was palpable everywhere.

So, like the girls, cast to the spatial margins of the here and now of white colonial habitation, I found my way to a motel that collected the condemned.[23] Much like Mary, Jane, who'd "been in the life for some time now," was careful not to tell me too much about her past. She was quick-witted and sarcastic, and she knew when, where, and how to get new clients, tips she passed on to the other girls when the regular supply of Johns dwindled. Mary spoke slowly, cautiously. She struck me as older, a different kind of wise that comes with watching things change without things ever really changing. And while she took the longest to "figure me out," as she would say, she was never hostile. Reserved and always observing, she was slow with details, almost like she was self-redacting and self-constructing at the same time. Betty, on the other hand, told me many things about herself, most of which conflicted, from the different names she used for herself to how many times she'd been to prison. All the girls, in their own ways, constantly obscured their origins. The modes of place-making that can turn anything (including or especially the motel parking lot) into a fugitive living room don't require the myth of origin stories, especially when origin stories of condemnation are not of your own making.

Despite Mary's initial suspicion that "Black people don't talk about the environment" when we sat together on the hood of my car with Clipse in the background, she began to ruminate, plumbing her own inventories and finding her past in proximity to waste: "I've been thinking about this question you asked. About Black people and waste? I guess I ain't never thought about it like this before but I've been thinking about how my family never had the truck come by. We never lived in that kind of place. When I was a kid, we used to have to take our garbage to the city dump. No one picked it up."

She lit another cigarette and handed me her lighter, an invitation to have one too. "Anyway, I remember when it closed, there was a big thing in the paper about it. You think that had something to do with racism? I bet it did. Man, I wish I remembered exactly what it was 'cuz I bet you anything it poisoned me. And my daddy too." She shifted a bit on the hood of the car to turn toward me. "I used to ask my dad why we had to take our garbage to the dump and others didn't." She lowered her voice to mimic her dad, putting on a slower, heavier Southern drawl: "'We have the opportunity to prove we care about things too. We don't need white people to keep our streets clean, we do it ourselves just fine, baby.'" She shifted her hips to face forward again, shaking her head. "I love my daddy, God rest his soul, but I don't want to keep the street clean; I'm tired."

Mary's family had initially lived in Norfolk, more specifically in Ghent, where this chapter began. Norfolk has been shaped by racist spatial violence. Jim Crow segregation was, among many things, an urban planning strategy. Undergirding Jim Crow's "separate but equal" strategy were racial colonial logics; separate accommodations protected against the potential of racial contamination, reinforcing white superiority through segregated public facilities such as parks, beaches, restaurants, businesses, and other leisure spaces. Norfolk's city council formalized residential racial segregation in 1914, enforcing a spatial color line with fines and fees. A series of legal restrictions on Black movement hardened the color line, including prohibitions on so-called miscegenation, compounding the belief that Black blood was contaminating. Segregation's spatial strategy was also, of course, economic, as are all spatial strategies under settler-colonial racial capitalism. After the formal end of slavery, Jim Crow recomposed slavery's racial-spatial logics of exploitation, expropriation, and accumulation: not just separate but *unequal* housing, jobs, infrastructure, and so on. What *Plessy v. Ferguson* established federally was implemented with alacrity in postbellum Southern cities, legalizing disinvestment in Black life.

Though exploitative renting practices, racial covenants, and racial violence characterize urban development in the 1900s, the emergence of residential security maps in 1933 and the Federal Housing Administration (FHA) in 1934 further entrenched spatialization as a way to ward off the imagined contamination of racial degeneration. These maps turned the racial character of a neighborhood into a calculation of mortgage lending risk; in other words, the "security" in the maps refers to the security of the *investment*, not to the security (or precarity) of life itself. In response to the Great Depression, Roosevelt's New Deal created the Home Owners' Loan Corporation and the FHA to stimulate mortgage lending. While white neighborhoods were graded "green" (the best places for financial investment), Black and racially mixed neighborhoods (with Black people) were graded "red" (indicating a *hazardous* investment). Redlining, as the quintessential mapping of hazardous risk, quickly became Jim Crow's preferred tactic in its arsenal of de jure segregation, and eventually it would wipe Black residents in Norfolk off the map.[24] The line from redlining to the toxic threshold is clear: If Blackness and Black places are already hazardous and risky investments, placing more risk there can't do any harm; in other words, there is no risk too great for Blackness to assimilate.

Turning the color line red mortgaged Black lives for white futures. Despite (or perhaps because of) 1954's *Brown v. Board* decision, white resistance to desegregation led to an intensification and a proliferation of the tools of racial containment. Despite the growing postwar pressure for new housing, the Norfolk Redevelopment and Housing Authority used FHA funds to buy and raze Black homes, leaving lots empty for years. By the 1970s, hundreds of Black residents had been forcibly removed from Norfolk, and in 1973, Mary's family was violently expelled from East Ghent when the neighborhood was seized by eminent domain. The lines of violence already carved by the disparities in landlord renting practices, city ordinances, and resistance to desegregation were etched deeper into the future schematic of the city. To be clear, this etching of racial and financial segregation was predicated on the transformation of Black people into social and environmental *hazards*.

Pushed out by racial capitalism's settler-colonial logics, Mary's family moved to where land was cheap, to what has become one of Suffolk's two Black zip codes. Across the United States, predominantly Black low-income neighborhoods report higher rates of asthma, low birth weight, and cancer, and higher incidences of lead poisoning—health outcomes common to living near LULUs, such as highways, plants, Superfund sites, and landfills. Southeastern Virginia is no different.[25] In 2007, the region's newspaper published an article documenting the high-risk lead exposure zip codes of the Hampton Roads, all of which were predominantly Black.[26] In Suffolk, the two Black zip codes are not only bisected by a privately operated freight railway and dotted with factories but literally surrounded by LULUs and unremediated toxified land. To the east is the Great Dismal Swamp, declared a wildlife refuge in 1974 after the extractive economies of the plantation, centuries of logging, construction dumping, and toxic leaks made the swamp's ecologies unpredictable; to the north is the currently operative regional landfill, owned by the Southeastern Public Service Authority (SPSA); and to the south is a Superfund site, a defunct SPSA landfill.[27]

"We went to that dump all the time and I don't know what the fuck was going on in there 'cuz there was a big think [sic] about it in the paper. I remember that. But also my daddy got sick right after and I just, I don't know . . ." Mary trailed off. She seemed to be simultaneously making connections, bringing different forms of knowing into proximity with one another, then becoming immediately skeptical of her own knowing.

"I mean, probably wasn't the dump exactly, right? Who even knows? People don't often live that long where I'm from. And we moved too, you know I grew up in Ghent? And, well, that shit is just gone. Seriously gone, just knocked the whole thing down. You know shit ain't right where we [Black people] live and I'm not talking about the people, I'm talking about *the shit that you can't see.*" In her hesitations, I was reminded of my own. Investigating the relationship between Blackness and waste in an anti-Black world is fraught with epistemological traps, including the desire to prove that the relationship is not born of pathology. Looking wistfully into the parking lot that was never full of cars but at night would fill with the girls and their stories, she affirmed something I know too: "It's like you just know something, just because of where you live. It's like where you live, or where *we* live, is not . . . natural. Do you know what I mean?"

The dump in Mary's story was SPSA's defunct landfill. Originally in operation from 1967 to 1985, it was a sixty-seven-acre unlined landfill that primarily managed Suffolk's municipal solid waste (and sometimes waste from the surrounding county). In 1983, SPSA decided to expand its operations. It began construction on a regional landfill for the seven cities of the Hampton Roads and started to plan for the closure of the Suffolk landfill.[28] While preparing for the closure, the city uncovered something it hadn't expected. According to Environmental Protection Agency (EPA) records, "The City discovered documentation indicating that several tons of debris that contained pesticides had been disposed of in the landfill in 1970."[29] The approximately twenty-seven to thirty tons of organophosphorus pesticides had been damaged when a fire broke out at a southern fertilizer company, the Dixie Guano Company.[30] First consulting with Virginia's Department of Waste Management, the state suggested that mixing the contaminants with lime would be enough to avert further environmental damage. The EPA's initial reports, however, showed elevated levels of Di-Syston (an organophosphate used in pesticides/insecticides) and arsenic. Thus, the EPA designated the landfill a national priority and placed it on the National Priorities List (NPL) in 1988.[31]

But the city contested the contamination of the site, arguing that the EPA's assessment of contaminants was speculative and that concerns about the poisoning of groundwater and the swamp were overblown.[32] The city, if held responsible, would have had to spend upward of $32 million to remediate toxification. Instead, the City of Suffolk hired its own consultants to determine the nature and extent of the contamination. The report, commissioned in 1992, backed up the city's position. No pesticides were found in the groundwater, surface water, or sediment, suggesting that the pesticides had degraded naturally over time. Though arsenic was cited as a contaminant in ground and surface water near the site, because the closest private wells were "upgradient," according to EPA and Department of Environmental Quality (DEQ) regulations, the level of risk did not warrant any remediation. As per EPA guidelines for monitoring Superfund sites, given that the pesticides and their derivatives had mysteriously disappeared between an assessment conducted by one group of scientists and an assessment conducted by another, all that was deemed environmentally "necessary" was the installation and periodic checking of groundwater monitoring wells. The EPA deleted the site from the NPL in 1995. And after the fourth and final five-year review in 2014, the matter was closed. It should be noted, however, that although lead was *not* a contaminant of concern, the investigation did find it.[33] In the poor Black parts of town, Suffolk was positioned as the threshold where risk and toxicity accumulate and are allowed to dissipate "naturally"—that is, through the environment absorbing, accumulating, and circulating toxicity until it can no longer be measured. Under the anti-Black calculations of the (settler-) colonial threshold theory of pollution, Black bodies and Blackened places are a surplus absorbing any and all matter of risk, not *people* entangled in land complexities, much less entanglements in need of attention (e.g., remediation).

While Mary never made explicit connections between her father's illness and the dump, her speculation unearthed how proximity to waste is produced by an anti-Black land strategy that shuttles Black dispossessed persons from one hazardous site to another. Clearing the hazard of Blackness from Norfolk to toxic zip codes in Suffolk shows how dispossession is an ecological condition. And this form of environmental management obscures how multiple directions of toxicity cannot be understood *merely* through acute sites of exposure. Instead, the "where[s]" of Mary's life not only are critical to *what* she knows but also mark *how* she knows, as they are the "excess" constantly wiped off the map.

The science of land use and the measurement of toxicants are the grammar of environmental reports. Yet they belie what it's like to live as the sink for pollution. "My brother was born different, small, you know? It was a shock to the family because we're all big and juicy," Mary said, laughing. "Not like you, girl," she teased. "You need to eat some more McDonald's!"

I laughed. "You know, I was a small baby too, don't even think I weighed six pounds. I was also really sick when I was young . . . have been sick my whole life, really." I told Mary about how I was diagnosed with leukemia at five years old, how I'd lived in similar neighborhoods and places with lead paint, how I've been diagnosed with multiple chronic autoimmune disorders, including a seizure disorder that emerged late in life. "Even now," I said, "I live in constant pain."

"Damn," she responded, shaking her head. "You know, you think getting an education and all that shit will mean you don't have to live with all these ailments, but it really don't. Guess there's no getting out of being Black, huh?" Though Blackness's relationship to geography and ecology might be disavowed, *being* Black is information enough to connect chronic pain to disabling environments, even if the causal directions are unclear. Toxic capture makes inhabiting the flesh chronically constructed as a "hazard" something you can't transcend. Hazardous conditions follow the construction of Black life *as* hazards. Toxic capture is material, and it follows.[34]

"He was still an infant when he passed; the doc said it was lead poisoning, said I probably had it too, but I was older and already getting into trouble at school." The doctor treated Mary's potential lead exposure as a problem of her body size, a pathological link that, as Black feminist historian of science Sabrina Strings writes, presumes the Black body is a priori diseased.[35] "They didn't care. All they said to me was to get more exercise. You believe that?" Pathologizing her fatness allowed the doctor to disavow the conditions of Mary's life, casting her outside the purview of public health and into the realm of personal responsibility. According to epidemiologists, there is no safe level of lead in blood: "The damage is permanent and untreatable."[36] Despite a wealth of scientific knowledge that, as early as 1976, proved that lead was a profound neurotoxin, "even at levels far below those considered dangerous,"[37] the guidelines of the Centers for Disease

Control and Prevention (CDC) for recommended action have been remarkably slow to lower the level of concern. As late as 2002, no state action was required unless levels fall within a range of 10–14 μg/dL and above. According to the CDC, blood lead levels (BLLs) less than 5 μg/dL "may not trigger a Department of Housing and Urban Development (HUD) environmental investigation. . . . Additionally, environmental investigations for BLLs that are 3.5–19 μg/dL vary based on jurisdictional requirements and available resources."[38] An environmental investigation (meaning a test of soil, air, home, and groundwater) is recommended at all blood lead levels, though its coverage by community grants, HUD, or health insurance and/or Medicaid is not a guarantee. Moreover, while the CDC's current recommendations for lead poisoning have adopted 2 μg/dL as a level of concern, it was not that long ago (1991) that 10 μg/dL was "established [as an] acceptable blood lead level . . . for children."[39] Caught up in the acceptable threshold of poison, Mary's flesh was pathologized and punished. Being poisoned became a lesson in the way the "excessive" body is captured. The doctor's advice to Mary was to better discipline her flesh, even though exercise is an unreasonable prescription for the lead exposure she could not have prevented.

"People in my family be getting cancer. Ain't no *love* for us." "Trust there's hate here." "If it's not the cops killing you, it's your food, your damn house, or that bad soil—ha!" Stringing together Mary's sentences makes all the clearer how anti-Blackness manages the environment: The use of Black people's bodies to absorb risk is inventoried by the individual who cannot trace the source of toxic capture but is condemned to the condition. Whether it is the Black body incorporated as the threshold of pollution, or Black life dislocated in service of city cleanup and rejuvenation, to follow trash is to run into Blackness. Which I did, time and time again—including my own. Mary's story is like so many I'd heard and lived before, such as my speculations about where I grew up and whether or not there is a connection between my litany of health problems and the environments in which I was struggling to breathe. But more than similarity, the likeness of the stories that echoed across my notebooks began to produce an unwieldy accumulation, an epistemic problem: my desire to prove that Mary's story was not *just* a story but empirical evidence about toxicity's truths.

In the years since leaving Virginia, I've returned to toxicological reports, to EPA Superfund maps, to independent- and government-commissioned reports, each time looking for a smoking gun, a piece of evidence that will break the story wide open. We are indeed caught in what M. Murphy has aptly articulated as a "chemical regime of living," in which the strategic production of knowledge about multiple exposures makes it difficult to define, isolate, or identify a "source." Moreover, as Murphy notes, when injurious molecular relations *are* acknowledged, "they tend to be posited as the acceptable contractual risks of laborers, or as the legitimate cost-beneficial risk to consumers."[40] While the categories of "laborer" and "consumer" point us to *some* of the sites of accumulating risk, they disavow the way that racial capitalism's spatial methods concentrate risk over and over again. Though the injurious molecular relations of racial capitalism are constantly and intentionally obscured, they also seem to be presumptively *acceptable to*, if not defined by, the "pathological," "poor," and "criminal" people considered hazardous to "the environment."

Toxic capture is an incalculable contract in which the (criminal and pathological) Black poor do not have to consent to being the threshold, nor is their positioning at the threshold ever a matter to be *measured*. In other words, when Blackness is positioned as the outside to value, no harm to Blackness *can be calculated*, there is no ledger in which it is tabulated; instead, it's inventoried by flesh.[41] As Murphy argues, this is why the ever-accumulating evidence that the infrastructures of everyday colonial habitation poison has had minimal political effect. The accumulated poison is accumulating where it "should" be, colluding with the racist belief that nonwhite people are determined by their environments. As with Mary's doctor's exhortation to exercise as a treatment for lead exposure, this line of thinking takes aim at poor peoples' "unhealthful" behaviors and never condemns (settler) colonial racial capitalism as the thing that needs to be changed.[42] This is not to undermine how communities contest their conditions, including through the grassroots science that drew attention to the ongoing Flint water crisis, the water pipelines in Newark, New Jersey, or the long fight over the Warren County landfill in North Carolina.[43] Instead, it is to draw attention to the way that toxic capture always attempts to undermine poor Black people's claims to environmental knowledge on the one hand and creates a threshold where environmental contestation *necessarily* challenges what we think of as "environmental" or "ecological" ethics on the other.

Black geographic knowledge is ecological knowledge. As such, I want to urge us to think alongside Katherine McKittrick, to shift "from studying [environmental] science to studying [ecological] ways of knowing."[44] From within Mary's inventory, the science of land use and the colonial measurements of toxicants obscure what she knows about the environment, including what she's "allowed" to know about her life, the land, and the relationship between ecology and Black flesh. Instead, *place* (and the emplacement of waste) is how she, and others like her, knows the threshold of harm. The threshold—the point at which toxicity is acceptable—is *material* even if unprovable, and "biocentrically induced accumulation by dispossession" is geographic and ecological punishment.[45] The carceral logics that produce the places to which "hazards" belong also inform how Black people (and, perhaps more specifically, criminalized Black people) *know the environment*. Punished all her life for her size, her weight, her flesh, her dispossession, and her behavior, Mary refuses to make herself small as an answer to toxic conditions: "Ain't nothin' wrong with *me*; something's wrong with this fucking world." If all "Black matters are spatial matters" and all spatial matters are ecological, then when Mary (re)writes her fleshiness into the world—"I am a goddamn goddess whose juicy ass is a gift"—she too becomes a Black ecological thinker.[46] Using her material knowledge to insist toxicity is not a problem within (flesh) but a violence that starts without and cleaves its way in, Mary sounds a call for something akin to abolition. With a different emphasis this time, she says it again: "Ain't *nothin'* wrong with me, girl, something's wrong with this world."

"RESPECTABLE" ENVIRONMENTS (OF SOCIAL DEATH)

Since the murder of George Floyd, "I can't breathe" has become a rallying cry for the Black Lives Matter movement. Within the violent conditions choking Black life, as a political refrain, "I can't breathe" announces how anti-Blackness is atmospheric. From police murder to acute pollution, the risks accrued to Black people are multiple, and they turn Black breathing into a political act—sometimes as resistance, sometimes as refusal, and sometimes as the quiet (even fugitive) histories of people breathing *anyway*. In the stranglehold of toxic capture, the archives of breathlessness carry the traces of stakeholders, regulatory apparatuses, scientific paradigms, and property relations. The struggle to breathe is marked by the ongoing effects of a "vested interest in whiteness."[47] As Darius's inventory devastatingly notes at the outset of this chapter, when it comes to the acceptable risks to Black life, at the threshold, risks accrue and no risk is too great for Blackness to assimilate.

In Portsmouth, known by tourists for its crime rates but by Black residents for its high rates of respiratory illness, the air is different: "It's thick, heavy even." Just across the waterway from the postindustrial port city is the Norfolk Naval Shipyard. Staring at the shipyard at the mouth of the James and Elizabeth Rivers, Darius continues, "The air hangs lower, like it gets stuck in my lungs." In just a few years, moments before the pandemic lockdowns begin, the US Navy will announce a proposal to build a new power plant that will galvanize opposition by Black Lives Matter activists and Portsmouth's chapter of the NAACP. But in 2015, meditating on the labor of Black breath, Darius's prescient description of the struggle to breathe predates COVID-19. Unknowing of the powerful stakeholders (the navy, the DEQ, the EPA, the waste industry) whose chokehold over the river would further entrench risks to Black residents in Portsmouth, Darius coughed as murky water lapped at the dock. And, still coughing, Darius lit a joint.

Inhaling, he said, "When you're there, it's like you can't catch your breath. Like you don't want it in your mouth. Ironic, I guess; n****s be holding their breath, only to have it choked out of them by cops. The shipyard is toxic. Everyone knows. And so is that plant. They say it's 'green,' but you go look at that shit and tell me what *clean* energy is?!" The plant to which Darius refers is a waste-to-energy (WTE) facility called Wheelabrator, a subsidiary of WMI. The facility was initially owned by SPSA, the region's waste-disposal cooperative, until debt and pressure from the private market threatened operations. WMI, one of the largest waste corporations in the United States, controls a significant portion of the waste flowing through Virginia. With contracts as far away as New York City and as close as Washington, DC, WMI's management of massive waste flows represents the changes that the privatization of the waste industry wrought. The importing of out-of-state waste (the management of waste across long distances) and its diversion to WTE facilities owned and operated by waste conglomerates produce chokeholds through which waste flows. Waste imports in Virginia peaked at 7.7 million tons the year Darius and I sat at the dock, an increase from the year before of over 500,000 tons.[48] Inside the conditions of toxic capture, where toxicity is multidirectional, unprovable, and the punishment of Blackened places, the 3.4 percent increase in out-of-state waste tonnage was barely noticeable. What Darius *did* find important to emphasize, which he repeated on an exhale, was, "That plant *ain't* green."

In the recent uptick of zero-waste campaigns, postindustrial towns such as Portsmouth have been heavily inundated by WTE facilities like

Wheelabrator. Zero-waste-to-landfill campaigns have promoted WTE plants as a technological fix to the waste accumulations endemic to racial capitalism. The term *zero waste* is, however, a mystification. Zero waste does not mean eliminating waste at the point of production; rather, it is an attempt to make waste "productive"—as energy or reclaimed/recyclable materials. It should be no surprise, then, that WTE plants are primarily owned and operated by private global waste conglomerations that cannot disinvest from the private market on which they rely. These facilities have retrenched market-based regulatory arrangements, whereby places of accumulated risk are turned into an exchange of health *for* capital.[49] The privatization of waste has secured a valuable commodity in which acceptable risk calculations are enrolled.[50] Diverting waste from landfills to create energy has, in many Black zip codes, inundated communities with the often invisible burden of particulate matter hanging in the air.[51]

The Virginia Asthma Coalition put out a Virginia Asthma Plan for 2011–16. The plan, written in 2010, was funded by the Virginia Department of Health (VDH) through the CDC and supported by the governor, Robert F. McDonnell. In his letter of support, McDonnell wrote, "The Virginia Asthma Plan urges those with asthma to be self-aware, make good choices, and manage their condition."[52] In complement, Virginia's health commissioner, Karen Remley, wrote, "Although it is a leading chronic health condition among children and adults, responsible for lower quality of life and undesirable health outcomes, asthma can be controlled."[53] While the plan suggests strengthening relationships between health-care providers and patients, standardizing asthma care, promoting community-based asthma educational programming, and pursuing legislative and nonlegislative policies, there is a startling lack of emphasis on the *systemic* environmental factors at play. The precise situations through which Black residents in Portsmouth are *made vulnerable* to respiratory disease are obscured by how asthma rates are "counted" as an aggregate percentage of residents. The focus is on documenting the diseased body, not identifying the *causes of* air poisoning and the harm to ecologies of which people and land are a part. This is most obvious in the way that asthma rates are reported.

The Virginia Asthma Plan estimates asthma rates by census tracts. In Portsmouth, asthma is reported to affect 7.9 percent of the population, while Norfolk asthma rates are among the highest in Virginia at 12 percent. But the data is aggregated and not broken down by race, neighborhood, or income. As feminist and antiracist Marxists argue, the spatialization of race, class, and gender produce uneven health outcomes: Socio-spatial processes determine not just exposure to the causes of ill health (on the job, in the home, and on the street) but uneven access to health care, capacity to move (exchange a toxic job for a less hazardous one, move from a lead-filled apartment to a safer one), and resources to change the material circumstances of one's (and one's community's) existence (prevent a toxic WTE plant from setting up shop down the block). But even reports that break down asthma rates by race, class, and gender, like the 2018 VDH *Asthma Burden Report*, cannot calculate the *compounding* factors of race, class, and gender, or how racism, classism, and sexism converge to determine exposure.[54] According to the 2018 VDH report, 11.4 percent of Black Virginians and 14.4 percent of the lowest income Virginians have asthma; women have higher asthma rates than men. The report does *not*, however, give data for low-income Black women; it would seem that the rates of asthma for poor Black women are not *worth* calculating (if the report did so, no doubt the rates would be higher).

But the problem with the data isn't merely one of missing calculations. It is that asthma-rate reporting—even at the "aggregate" level of the population—still treats asthma as a problem of (a collection of) individual ill bodies, not a problem of the toxic conditions of exposure. The reports leave unanswered why and how particulate matter comes to hang in the air of poor Black neighborhoods. The reports form a web of epistemological punishment, where evidence of environmental racism is turned into a weapon berating poor Black people for not preventing their own poisoning. While the Virginia Asthma Plan might urge people to "make good choices" in order to lower their asthma risk, the struggle to breathe in Portsmouth is not a choice; it is the outcome of compounding risks that accrue at the threshold of toxic capture.

By 2020, when the navy proposed *another* power plant, Wheelabrator stood as yet another hazard to Black life. The navy's proposal for a new plant reframed the once "dependable, environmentally safe" Wheelabrator facility as an environmental problem.[55] Though Wheelabrator's website proclaims how much waste is diverted to the facility, turned into energy, and sold on the grid, the navy's proposal to build a plant inside the shipyard purported to "supply the facility with a cleaner and more reliable source of energy."[56] Black residents opposed the plan, arguing that the plant would compound their respiratory problems and that the undue burden of pollution would further imperil Black lives. Moreover, they argued, "There's no guarantee that the Wheelabrator plant will close after the new plant opens, which could mean the impacts would be cumulative."[57]

Community opposition was dismissed by the DEQ, whose statistical models of the navy's plan argued that it would add "little pollution to the surrounding air," noting that "the existing air quality is good enough, in part because of the presence of ocean breezes to absorb the plant's additional emissions." James Boyd, the local NAACP chapter president, disagreed: "Just because it's an acceptable level of [pollution], doesn't mean it's a humane level."[58] The threshold theory of pollution's use of the Black body structures the above debate: Additional pollution is always "acceptable" when the potential sink for it is Black. While the DEQ and the navy might deny it, the disavowed use of the Black body as the threshold is evident. The entitlement to pollute the water and the breeze is a colonial resource for disposal that is reinforced by the captive (and already diseased) Black population in Portsmouth. The environment is rendered available because Black lives are worth *less*. Moreover, even within the settler-capitalist's scientific projection of calculable pollution, no one can name what will absorb the "acceptable" projected 2 percent increase in soot if and when the breeze changes or water levels rise.

This threshold's colonial relationship with matter, including the mattering of Black *flesh*, makes risk *accrue*. Boyd's critique, that "they're looking at the economic impact and benefit it might have for the shipyard" and not people's lives, is further bolstered by the fact that the shipyard is already a Superfund site.[59] And it's not Portsmouth's only hazard. All Black residents in Portsmouth live within fifteen miles of a Superfund site, and according to environmental reporter Jeremy Cox, "Within a 2-mile radius of the project, 70% of the residents are members of a minority."[60] Near the shipyard, there are eleven fuel-storage locations, four of which are Superfund sites, two of which are on the National Priorities List. And there are countless others.[61] Despite all the evidence suggesting that a new plant would further jeopardize Black residents' health, in 2020 Virginia's Air Pollution Control Board accepted the proposal in a five-to-one vote.[62]

The pooling of risk is infrastructural, historical, and racialized; the value of whiteness as property hides toxic capture in the data. Statistical models project entitlement to the totality of our environments, presuming that ecological relations are obvious, stable, and isolatable. As the climate warms and weather patterns change, there is no guarantee that current projections accurately describe future weather patterns or future ecologies' capacities to "assimilate" toxicity. The notion that there is some level of acceptable pollution guides how risk is calculated, presuming that calculations of risk to ecological health are "unmarked" by power relations and unaffected by risks to land, climate, water, atmosphere, and marine life. The anti-Indigenous threshold theory of pollution and its attendant anti-Black socio-spatial relations of toxic capture turn geography and geology—the *where* and *what* of place—into racialized conditions of ecological violence that allow toxicity to pool and secure the coordination of a waste market.

As a port city, Portsmouth is an infrastructural chokehold, part of what Deborah Cowen might call a "corridor of power."[63] If "acceptable risk" is in part calculated by ignoring the spatialization of race, then corridors of power materialize environmental control through the disavowal of toxic capture. For this reason, it's important to understand not only that Wheelabrator is a facility that is hazardous to Black life but also that it is caught up in the management of waste. As a WTE facility, Wheelabrator affects the profitability of the region's waste infrastructure. Thus, it is also caught up in the politics of the public regional waste infrastructure, operated by SPSA.

The privatization of waste has pit industry giants like WMI against public waste management systems like SPSA. As a result, local waste management infrastructures have had to adapt to the private market's demands, where staying afloat means competing for the "commodity" (i.e., waste). But racial capitalism's death-dealing practices defining "acceptable risk" *also* define how market antagonisms will resolve: The colonial entitlement to land and the fungibility of Blackness determine where toxic capture is most profitable.

The coordinated infrastructures at the mouth of the James River—the shipyard, Wheelabrator, and the proposed plant—embed the fungibility of Blackness and the colonial entitlement to land. Before returning to how these plans, and the reports generated around them, trap Black residents as if they are *a hazard of their own making*, I want to turn to some of SPSA's failed plans for expansion. I do this because the failed projects reveal two things. First, they make visible local waste management's relationship to the privatization of waste. Second, they show how, in trying to resolve market antagonisms, SPSA relied on the *elsewhere* of a Black threshold. These failures show us that toxic capture is not incidental but materializes colonial violence by infrastructural design.

By the 1990s, SPSA was in spatial and financial crisis. This crisis was due to a combination of factors: the public bond that financed SPSA's construction in 1976, the cost-cutting pressure put on the public cooperative because of the private waste-disposal market, and the demand to take in more waste to finance landfill expansion. Without money to expand its waste-disposal operations, the regional landfill was on the precipice of environmental crisis. SPSA's bond rating was in jeopardy, threatening to shut down landfill operations. But SPSA was necessarily running on debt. After trying and failing to borrow more than $22 million to finance improvements "for environmental health and safety reasons," SPSA began a series of attempts to partner with private waste companies to finance expansion.[64] While all these attempts failed, and SPSA's spatial "fix" would come from a regulatory work-around in a Black zip code in Suffolk, these failures are as telling as SPSA's eventual success.

In 2006, SPSA attempted to build a trash port in Portsmouth. (Yes, that same Portsmouth where the navy proposed its new plant and where all Black residents live close to Superfund sites.) Envisioning a stream of waste barges to the Blackened port city, SPSA courted waste megacorporation Covanta, a company that only a year before had been forced to file Chapter 11 bankruptcy after settling a number of air pollution lawsuits across the country. The partnership would've been mutually beneficial. For Covanta, partnering with SPSA would have helped them gain a foothold in New York City's waste, a commodity that had become a turf war for the industry since 1996, when Giuliani announced that he was working toward the closure of the city's Fresh Kills landfill. For SPSA, it would have established its outpost in the private market. Had this trash port been built, the public-private partnership would've compounded the environmental risks that already accrue in Portsmouth and accumulate, punitively, near and atop sites of Black history. Not only would the proposed trash port have reinforced the capture of environments—Black bodies, the water, the air, the breeze—but it would've been located near the Emanuel African Methodist Episcopal Church, once part of the Underground Railroad—a place that, because of low land values surrounding the church, had already made it hard to claim a place in history.[65] While urban scholars have written about how the historic districting is used to insulate private-property owners from the development of low-income housing and the placement of LULUs, less has been written about how low land values overwrite Black history.[66] Unsurprisingly, Black residents in Portsmouth opposed the plan, siting the number of Superfund sites near their homes, including the shipyard. However, it was not their lives that successfully sunk the partnership; instead, it was the town of Chesapeake, with a higher percentage of white residents, that successfully argued the port would impede the view of a condo development project south of the river.[67]

Next, SPSA floated a partnership with Black Bear LLC to build a mega landfill in the wetlands of Camden, North Carolina. The proposed mega landfill was to be sited near the southeastern part of the Great Dismal Swamp, a place long nurturing of Black and Indigenous resistance.[68] In 2012, new regulations already stipulated that landfills could not be built within a five-mile radius of a wildlife refuge. But the mega-landfill project was strongly opposed by the 90 percent white city of Camden and was scrapped. In each case, these public-private partnerships were blocked because of the effects they would have had on white property. That white wealth could block these proposals is not a surprise, but it does evidence the way that NIMBY (not in my backyard) is not only a strategy for insulating white life but a strategy that is effective because white lives are considered more valuable. Had either of these projects been successful, they would have created an infrastructural link that expanded SPSA's disposal capacity. However, I want to draw attention to these (failed) sites of expansion because they articulate a pernicious relationship between race and waste infrastructure's capacity to choke the meaning of *place*.

The NIMBY politics that quashed SPSA's private ventures protected the value and meaning of white property above all else. White property defines the value of *here*, and Black meanings generated elsewhere—the meanings of *place*—are easily written over. In poor neighborhoods like Portsmouth, historical landmarks are hard to find.[69] These "geographies of self-reliance" are also histories of the struggle to breathe.[70] I note these fugitive geographies—the Underground Railroad, the Emanuel African Methodist Episcopal Church, the Great Dismal Swamp—laden with other ways of valuing life and land, because their contradictory meanings to SPSA's strategic partnerships and to Black history mark how infrastructural plans project the *elsewhere* of toxic capture.

All geographies and ecologies that mark the struggle to breathe mark the struggle to *be*. In the elsewhere of a poor Black backyard, accumulated risks disavow history and other ways of generating meaning, other ways of being. "The ecology of injustice that structures urban life" is itself structured by waste infrastructures.[71] And even failed projects reveal how chokeholds are produced by design. The privatization of waste and the market competition it produces require infrastructures entitled to land and bodies. Toxic capture is always and never there, and questions the utility of interventions that discipline the behavior of people at/as the threshold. What does it mean to act "environmentally" in the poor Black backyard to which all risk is acceptable? When the mandate to be environmental is not a critique of how infrastructures support colonial living by accumulating risk to the Black and Indigenous other, it becomes the demand to live a respectable life of social death.

OUR ENVIRONMENTS ARE A PRISON CELL

I had watched Darius roll joints before, but I was particularly impressed by how, even sitting in the breeze at the river's mouth, he managed to get every last bit of weed in the roll. Expertly, he rolled tightly, and reading the changing wind, he waited to spark the joint until a gust was in his favor. "How can we be, if we don't have a right to breathe?" he said quietly. "Our environments are a prison cell. It's madness," he whispered softly into the breeze, where his sound was swallowed by the water.

In the opening passages of the evocative *How to Go Mad Without Losing Your Mind*, La Marr Jurelle Bruce writes us into the "unimaginable scene" of the slave ship to "launch a study of radical imagination" that centers mad Blackness and Black madness: "What vertigo does a body undergo, caught between treacherous waters below and treacherous captors above, with 'nowhere' outside?"[72] The violence above and the violence below structure the impossible conditions of breathing in the hold. As Willie Jamaal Wright insists, *as above, so below*. Thinking racism *as* environmental (as opposed to traditional notions of environmental racism), Wright argues that the "above" of state-sanctioned violence and the "below . . . of toxic waste in . . . soils and waterways" produces social death. In these conditions of toxic capture, Wright reminds us that Black *being* is also a critique that gazes back and "engenders the emancipation of the [Black] onlooker" and perhaps the madness of the Black ecologist.[73]

There is a kind of madness to environmental debates. Some environmental justice scholars still debate which came first: the noxious facilities or the minorities? Economists prioritize cheap land as an explanation for toxic siting decisions, never pausing to consider how racial violence makes land "cheap." The circular reasoning perpetually misses the mark: anti-Blackness is a(n) (anti-Indigenous sovereignty) land use strategy protecting the here and now of white propertied life by condemning Blackened environments and more-than-human life to the toxic hold. Racial zoning, redlining, and restrictive racial covenants make racism material, inscribing race into jurisdictional boundaries, plans for development, and zones for land "use." High rates of respiratory disease and the immanent capture produced by infrastructural plans are intimately entangled in daily breath, as are the risk calculations that presume land (and water) *should* assimilate pollution.[74] Yet the inability to think about the multidirectionality of risk, to understand anti-Blackness as *the weather* and (settler) colonialism's daily violence as atmospheric, blames poor Black communities for their own ill health.

In Portsmouth, public health narratives continue to belie the toxic capture of Black life, but living while condemned engenders different strategies to make toxicity "visible." Let's return to the run-up to the 2020 Air Pollution Control Board's vote on the navy's proposed clean energy plant, to be located on an existing Superfund site—the shipyard—near which 70 percent of Portsmouth's Black community lived. Pushing back on the navy's, the DEQ's and the Air Pollution Control Board's conclusions that the plant would not burden any particular community with pollution, Black activists began citing yet another asthma document, this time a 2017 VDH report. The report did not assess health outcomes based on demographic data, only by census tract, but Black organizers strategically used the data against itself, engendering a *Black* way of looking.[75]

For residents who *know* without data, figuring out how to reveal where race and class are being erased means looking at space. Because of residential segregation, census tracts can be used as a proxy measure for Blackness. As previously mentioned, since 1990 the spatialization of race and income in Portsmouth has changed very little, so focusing on the specific census tracts closest to the shipyard allowed Black activists to re-present the 2017 data as revealing disproportionate burdens to Black life. Though dismissed, residents gazed back at the shipyard reframing the data in state asthma reports to demand an accounting.

The VDH report miscounting asthma's effect on Black residents also (mis)names asthma's *solution.* The interventions recommended in the report are interventions into people's behavior—primarily smoking bans, healthy eating, and physical activity—without addressing racism's environmental toll. The report disavows the production of toxic capture and instead seeks to police, punish, and regulate how to be at the threshold. Simply put, the report ignores the *structural* determinants of health, a frame that would require the VDH to shift their analysis from how the pathologized behave to how *capital* behaves. Fundamentally, recommendations targeting behavior occlude how toxicity is spatialized as a land-use strategy for racial capitalism and how environmental condemnation is a form of capture that makes Black *being* a struggle. The recommendations target Black sociality and the reproduction of it (including through pleasure, like smoking cigarettes or weed or drinking soda) as pathological behavior, casting poor Black people as the cause of their own toxification.

And pathologizing poor Black people's behavior remains a standard response to environmental problems, to return to the maddening circularity of white environmental reason. Take, for example, Darius lighting a joint within the chokehold of toxic capture. You, dear reader, very well might prickle at this, given his vulnerability to respiratory disease. You might admonish him that smoking *anything* is a bad choice for his health. But quitting smoking won't stop the breeze from carrying particulate matter from the WTE facility to Darius's neighborhood, just like changes to *his* behavior won't eliminate how ecological risks accrue to Black life. (Changing how ecological risks accrue to Black life would require changing the behavior of settler-capitalists and the state acting in their interest.) Instead, I implore you to ask yourself, What desires *should* Darius have, when his toxic capture is always coming?

HOOD RATS AT THE THRESHOLD

Toxic capture makes Black living and the reproduction of Black life into sites of ecological punishment.[76] This tension makes for an impossible paradox in which the struggle to be *is* the struggle to breathe, and yet *how* you are is not considered ecological at all. The mad-inducing traps are material as much as they are ideological, and they put an impossible burden on Black existence to resolve contradictions made from without that cleave their way in. What does it mean for Black people in Suffolk or Portsmouth to be "environmental," to make good "healthy" choices, when the toxicity generated from without is always forced into Blackened places and flesh? What kind of choice is a life without pleasure, a respectable life of social death? Perhaps respectability itself is a type of mad-Black response. Perhaps it is a way to quell Black madness. Either way, sometimes Black people try to resolve the contradictory demands on Black existence by blaming Black people themselves:

> That's why it's also my job [working with incarcerated and formerly incarcerated youth] to teach these boys how to be men, to show them what respect looks like. I tell them every day, "You gotta clean up your behavior, pick up after yourself, have pride in your street to have pride in yourself." You know, it's like you treat people in the hood like animals, ask them to live like animals; let me tell you something, these black boys will act like *animals*! You got to teach 'em how to respect things but especially, which women to stay away from, you know what I mean? Trashy women, you know, hood rats. Like the women at your motel. They trap these boys in a hellhole getting pregnant and once you have a kid, you're stuck.

Not unlike the state-generated health reports that suggest community behavior is to blame, Darius identifies a "community" problem too: the animality of Blackness and the particular problem of poor Black women's deviant genders.

Darius's strain of respectability is telling of how gender and class are produced as sites of Black ecological struggle. Black women, especially when poor, have a particular relationship to the punishment of toxic capture—a reminder, too, that capture is *gendered*. Ironically, for the same reasons that toxic capture is a mode of transmogrifying Blackness and ecology, the fugitive (criminal) politics of being fungible are also construed as animal.[77] As the litmus of a Euro-descendant genre of human, Darius calls upon available descriptions of Blackness as animal and a problem of Blackened environments. However, Zakiyyah Jackson argues that there are other genres that house alternative modes than that required of the propertied man (i.e., *becoming* human)—a genre that perhaps Mary espouses herself. While "Ain't nothin' wrong with me, girl, something's wrong with this world" speaks to the condition that starts without and cleaves its way in, it also speaks to the possible alternatives practiced by "unrespectable" Black women.

The description "hood rat"/"trashy" woman is both knowledge *about* the Blackening of environments and an alternative standpoint from which to see them. While for Darius this knowledge is mobilized as an uplift strategy for young Black men—a lifting out of the dirt of poverty, of incarceration, of the pathological Black family, and thus, perhaps, toxic capture—his prescriptions for Black boys' behavior depend on the same contradictions that differentially surround Black women. Moreover, Darius's prescriptions mark the complex ways that Black women experience the toxicological threshold as not just a problem of behavior but a fundamental failure to be *woman*. The dirtiness of dispossession, where home is made on the move, marks poor Black women as fundamentally out of order, progenitors of filth, and a sign of contamination. That Darius calls upon "the hood rat" to mark how Black boys are captured speaks to how Black women are impossibly positioned as unable to produce a healthful home and an essential impediment to lifting Black boys out of "the mire."[78] As a potent symbol of the reproduction of Black pathology, the Black woman, especially when poor, is perhaps something more than dirty: fundamentally unnatural. The Black woman, as Betty liked to describe herself, might be "something of a fiction."

"It is dirty here, you know? But it's also home." I smiled and nodded at Darius, remembering the cramped places where I learned to think and tried to grow. "Of course, it would be nice to have some trees around here, breathe some other kind of air, but it's what we got. I guess it's kinda funny to call a n****'s house dirty when n****s be given mud to make a house. Home is home, you feel me?" For Darius, poor Black women can trap a Black man into a punishing existence—an existence that at times he articulates as a problem without. But sometimes, in the punishing demands on Black existence, he reaches for the logics that disfigure Black women. My point here is to remember that "there are lessons to learn," to evoke La Marr Jurelle Bruce again, "from those who make homeland in wasteland . . . mad black worlds to make that rise from a ship's wake, and questions that refuse answers but rouse movements," including the joy that emerges from the self-manumission of moving, even though captured at/ as the toxic threshold.[79]

Though Darius might've described the motel, offset by the Holiday Inn, as *nothing special* if not just *waste*, it was both a site of neglect and a refuge of collective care. The sounds of everyday Black women's poverty remade homeland in wasteland. Spilling through the walls of individuated motel rooms, remaking the interior a collective one, this was a place of knowledge that incorporated the shoddy materials through which the girls made a transient home. Sometimes the women would knock to each other through the walls, but most often they would yell: "You in there? You gotta shower cap I could borrow?" The walls were thin and they could be used to communicate, no need for a proper door: "Hey, Imma run to the 7-Eleven, you need anything?"; or a form of incorporation, inviting me in as kin, "Ask baby girl next door if she need something too"; or a way of forging mutual aid through the walls, "You all right in there, I'll come in an' fuck that white man up if he touch you wrong"; or a fugitive plan, "Tell baby girl to put on that music, I want to get liiit tonight! I'll hit her back after I trick, she know where to meet me right? By the dumpster, make sure she know the one!" The sounds of Black women's lives spilled through and out. White families would leave just as quickly as they arrived, in fear of the unruly place where the sounds of Black women aren't easily contained.

Living while dispossessed is improper; propriety, after all, flows from ownership, not the other way around. At the motel, where "recycling" is not distinct from stealing, rewriting matter was a refusal to be respectable. One night in the parking lot, where the girls had turned up my music, Mary, referring to a white male client who had just left, said, "Look, I don't care what any of these fuckers think of me." Though I didn't hear it myself, as the man left, the others heard him call Mary a piece of trash. "It's not my *job* to care." Listing out her labor and counting out her jobs on her fingers, she said, "My job is to fuck, get money and provide for my girls, my family, myself." The volume on my car radio was as high as it would go, and Nicki Minaj blasted while they danced in their nighties before getting dressed to go trick. Taking me aside to whisper in my ear, Mary said quietly, "Listen, I know who I am, and I know a thing or two about how much these men like their trash. So, when his eyes were rolling back in his head, I stole his motel key, so I can stay here a few extra days." Mary winked and turned back toward the women. And using the same elision that made "trash" and "Black woman" an object of degeneration, an animal or a subhuman being of gendered disgust, she sung out the illicit pleasure of moving, yelling, dancing, and fucking for money. She yelled while dancing, "I might be a hood rat, but *you know* I could get it, 'cuz I DO!"—a crescendo to Nicki Minaj's lyrics, "Dem a wine up dem waist, dem a pat the pum pum," shaking her ass, flashing the stolen motel key.

INFRASTRUCTURE

Virginia is the second largest importer of waste in the nation, second only to Pennsylvania. In 2003, "the Commonwealth took in 6.6 million tons of out-of-state waste . . . , a 22 percent increase over 2002's 5.4 million tons." It was the largest percentage increase since 1998, when Virginia began tabulating waste-import statistics, which was just one year after Giuliani signed a waste contract that sent the first shipment of New York City's waste on its long journey south. Environmental officials like Virginia Air and Waste Policy Manager Melissa Porterfield argued that the increase in out-of-state waste imports to Virginia may have had to do with the fact that the state had no tipping fee or tax on waste imports; still, she acknowledged that low fees alone cannot account for "why Virginia is such an attractive disposal point." Perhaps it is the facility's location (near major highways that run up and down the East Coast, not to mention waterways that make it easy to barge), perhaps it is the industry deals struck with host communities to keep waste volumes high and revenue up, or perhaps it is arbitrary. According to the director of environmental programs for the National Solid Wastes Management Association: "While it is important to keep your costs down, the difference between taking waste to Pennsylvania or Virginia isn't that big," sometimes it's about the closest highway exit. "That's the problem with interstate [waste transport increases]. There's not a single cause."[1]

There are multiple ways to describe why Virginia has become such an attractive place for the management of waste. It is true that interstate waste hauling moves across long distances, requiring reliable routes from point A to point B. Interstate 95 runs from Maine to Florida. Anyone who has driven along this route is likely familiar with the experience of being sandwiched between Mack trucks, hoping to pass one only to be sandwiched yet again. Sometimes these trucks are transporting waste. But location isn't only about the highway and its accessibility for long-distance travel; it's also about how racism shapes what is and isn't profitable, who lives where, and whose lives produce "value."

Poor communities of color host the majority of landfills that receive out-of-state waste. Waste facilities, such as landfills and plants, sign "good host" or "good neighbor" agreements that stipulate how the facility will offset the community's cost for hosting pollution. Depending on the facility, this can be anything from fixing potholes and repaving roads in disinvested communities to no or reduced tipping fees for local waste, as a trade-off for the long-term health effects that come from living near waste infrastructure. If needing streets repaved, potholes filled in, or "broken windows" fixed is part of what makes one a good host, then, "host" is defined by the racial terms that govern America.

It seems a strange coincidence that Virginia, the first established colony and home to the first plantation, has become a place for the management of waste. Or perhaps it is evidence of what Katherine McKittrick calls a plantation logic, the naturalized racial dispossession set forth by slavery's geospatial protype. At once a meaningful blueprint for development and a "juridico-economic architecture" for Blackness's structural relationship to native homelands, the plantation secured settler theft by chaining enslaved labor to the land.[2] But as McKittrick carefully notes, this arrangement of white sovereignty and Black labor "naturalized a plantation logic that anticipated (but did not twin)" how dispossession shapes Black life in the present.[3] Perhaps this logic of dispossession—that long shadow that slavery cast over its descendants—is evidence that Blackness infrastructurally shapes the management of waste.[4]

Thinking of the landfill as an outcome of a plantation logic reveals "the extent to which a limited understanding of scale," including a limited understanding of the scale at which waste is managed "is tied to a narrow conception of racism."[5] As Laura Pulido argues, racism is constitutive of the spatial scales that connect places to each other: The spatial violence of racism inheres in the way "places are produced by other places."[6] While environmental representatives and waste industry commentators debate the cause of the increase in waste imports to Virginia, the accumulating reality of waste is structurally enforced by good neighbor agreement contracts and the appropriation of land and poor people of color's bodies to host pollution as infrastructure. Tying together the history of suburbanization with the land-use practices that concentrate environmental hazards in Latinx neighborhoods in Los Angeles, Pulido argues that locally unwanted land uses do not just participate in producing the acute pollution or toxification of the racialized "inner city" but also stabilize the quality of life that white suburban residents demand. The ability to distance oneself from the concentration of hazards is structurally bound to the toxic enclosure of someone else. Together, the siting of hazards, which requires the planning of zoned districts of land for "use," and the production of property values that stabilize the economic use for "undesirable" land rely on a form of habitation that chains some to toxified lands.[7] The coordination of urban planning and the regulation and management of waste, produces predetermined outcomes on the one hand, and on the other, sediments the geographies anticipated by the plantation.

Pollution is a form of racial violence that Max Liboiron reminds us is deeply tied to the settler's presumptive "access to land."[8] Even the language of "land use" and urban planning's concept of "zones" occlude how colonial expropriation becomes a scientific if not moral duty of a rational state's use of our collective environments. Mapping "use" across ecotones, lives, and the more-than-human world introduces capital's cartographic demands to ecologic relations, requiring the material world to be suited for parcels. The lines of property run violently across the biosphere, through Indigenous ontologies of becoming with land and through Black lives in the wake of once being property themselves.[9] Black living is contradictorily and differently tied to land: as a tool of labor or as an ever-available "environmental" problem to be transmogrified in pursuit of it. The plantation logic that makes settler-colonial habitation possible is toxifying: Alongside access to land is the presumptive access to a captive Black body (to *host* the hazards of capital's own making). To put it another way, the toxic relations of confinement for some produce the place and time of another.

Prison abolitionists conceptualize incarceration's role in daily life similarly: The incarceration of some rationalizes the freedom of others. The prison, Ruth Wilson Gilmore and Angela Davis argue, is justified by a social and economic apparatus that participates in and upholds its maintenance.[10] The prison, according to Mariame Kaba, *teaches* us how to see a criminal, how to feel about the law and those who "break" it.[11] Judah Schept's work shows how prisons teach us to see people as landscape, creating a carceral economy of language through which commonsense descriptions of poverty confine the poor.[12] The prison teaches us to defer to the authority of the law to define crime and to see it as a necessary solution to large-scale and interpersonal harm.[13] It teaches us to desire police protection and to define safety as something predicated on security, surveillance, and invasion.[14] The prison naturalizes the logic of the plantation, and it does so by walling off life rendered disposable, to which any number of violences are excusable. The prison lulls us into an impoverished imagination of responses to social problems, and mystifies its use for managing "surplus" land and life as a "rehabilitative" apparatus that keeps us safe. But jails and prisons don't so much keep *us* safe as much as they keep *property* safe from other ways of inhabiting—including the Indigenous Land Back movement, reparations, and other ways of living that do not seek ownership over land.[15]

What if we imagined the landfill similarly. As a waste infrastructure to which we imagine the designated end of production and consumption, the landfill is both material site (the theft of land mystified by the science of land use) and thoroughly socialized (by the relations of geographic dominion). The landfill teaches: It naturalizes the desire to be distant from waste, to absent oneself from its accumulating reality, including those things that seem to gather around it (pests, the poor), and to fear its accumulation. Within this spatial arrangement—to which "away" marks what Malcom Ferdinand calls "a colonial inhabitation"—proximity to waste is axiomatically dangerous.[16] The "out of sight, out of mind" place of the landfill normalizes the segregation of places, an inheritance from the Progressive Era's development of zones, in which parcels of land have discrete functions. Similar to the way the prison can "fix" a crisis of land value for capitalism's racial geographies, zoning "fixes" the problem of hazards for racial capitalism's degradation of land, people, and more-than-human life.[17]

Spatial violence is ecological violence. The lines that parcel the earth follow directly from the belief that property is not a violence to the earth. And the landfill protects spatial violence through what it walls off *for whom*. The landfill's "absential" technique extends beyond the acute management of hazards to the social relations of cleanliness it fosters.[18] As a form of land use, it is tightly woven into the metrics of (de)value. Protecting property means not only protecting it from its wasteful accumulations but also protecting it from how those accumulations might *devalue* a propertied form of life. Most familiar in the language of "bad neighborhoods," *dirtiness* is an evaluation of land not yet living up to its potential, tacitly authorizing spatial violence in the evaluative modes that emerge around the (propertied) calculus of what is "not yet" white property. From redlining to gentrification to the policing of poor neighborhoods, the response to "dirtiness" is often spatial violence. As if it were the poor and criminalized whose environmental politics need improvement, the dislocation and displacement of those supposedly unclean inhabitants are signs of neighborhood "rejuvenation." Distinct from the prison in the sense that it is not the place where the criminal is held but perhaps connected in its relationship to producing the place from which the criminal is born, the landfill teaches us what to do with discards (including people discarded by capitalism): make them disappear.

The imagination of the landfill is critical to our contemporary environmental politics. We both abhor it and need it. It is a real place that is materially unruly, yet it is an unruly fantasy of rationally engineered order and complete capture, if not disappearance. The landfill is not a place of elimination; in fact, it is always producing new material realities that need to be addressed. The practice of landfilling waste produces leachate and methane. Landfills belch and excrete substances that need to be treated, captured, contained, and treated again. Leachate needs to be treated in order to protect water tables deep underground and marine life from concentrated metals. And the processes by which these flows of chemicals need to be carefully coordinated multiply with increasing accumulations of waste.

The ever-expanding need for landfills has shifted environmental activists' focus from the elimination of landfills to the promotion of circular economies. One such trend is that of campaigns for zero waste. In 2015, then-Mayor Bill de Blasio pledged to make New York a zero-waste city by 2030. However, zero-waste campaigns don't imagine a future where there is zero waste at the point of production or at the point of consumption; rather, they rely on recycling plants (which often produce or divert waste to landfills) and new waste-to-energy (WTE) technologies that attempt to turn a problem (waste) into a resource (energy).

In his attempt to deliver on his zero-waste pledge, de Blasio turned to a WTE plant in New Jersey owned by Covanta, a waste company with a notorious reputation for polluting and toxic dumping in communities of color. This is not, however, a story specific to Covanta. WTE companies are invested in so-called zero-waste solutions because they do not eliminate waste as a source of profit but merely redirect its flow. After all, waste corporations need waste; waste is their market. The rhetoric of "zero waste," much like the campaigns that have popped up in Seattle and San Francisco, imply that we can design our way out of racial capitalism's toxic accumulations. Ruha Benjamin's work, however, teaches us that you can't design the racism out of technology. Instead, she encourages us to rethink the desire for technology to fix social and political problems.[19] Much like the critical work on prisons previously surveyed points out, the social and political problems constitutive of racial capitalism cannot be solved by racial capitalism itself.

So, is Blackness infrastructural, providing the spatial and material justification for acute toxification, or does waste infrastructure manage Black life? Waste is part of "the arithmetic" of skewed life chances and the ongoing "vulnerability to premature [Black] death."[20] The landfill does more than describe the specific site of waste's disposal; it is critical to how we see what is and isn't *dirty*. It also reveals how the persistent assumption of land for use is linked to the disposability of Black lives. Because the landfill is the place where waste *ought to be*, it is productive of the evaluative schemas that emerge to protect property from the material and human hazards to it. In other words, the landfill is also ideological. It's critical to what we think is possible, but it's also critical to what we're unwilling to reckon with. Waste ain't going nowhere.

2 Becoming Fill

> I must be the bridge to nowhere.
> — RUSHIN, "THE BRIDGE POEM"

As I followed Jane out the door and into the motel parking lot, she snatched the cigarettes out of the pocket of my hooded sweatshirt. She lit two, threw the pack back to me, and handed me a smoke. The girls had already arranged the parking lot for us to gather and I could hear some of the others walking back from the 7-Eleven in their dollar-store flip-flops in the distance. There was little in the way of public space, and sandwiched as we were between highways, also little in the way of sanctioned social places. The woman at the front desk knew some of the women well and didn't seem to mind our gathering. Other than my strange arrival (that these women bring into question), the crowd was pretty consistent: There were the sex workers who temporarily lived in and out of the motel, the clients who would sometimes spend the night, and every now and then—though they wouldn't stay long when they encountered the "seedy" element of sex workers and drugs—young military families who couldn't afford the rising rents of Virginia Beach housing. Nestled out of sight a little ways behind the more "upscale" Holiday Inn, the parking lot—like the alleyway, the corner, and behind the dumpster—offered an interstitial corridor for people set aflight by the production of placelessness.[1] The place of being surplus to and for land.

"You ever been to Suffolk?" I shook my head as we made our way to the lot. "I think you'd find it pretty interesting." Jane struck me as younger than the other women. She asked me countless times how old I was, constantly in disbelief that I was older than eighteen. But each time I answered, she'd say, "*Black don't crack!* That's why I ain't never tell no one my real age!" So, I never bothered asking. "It's a weird place, Suffolk. Not much there. But maybe you'd like to walk the trail in the swamp. You seem like someone who might like that." I got the feeling that Jane understood my research to be more in the vein of "saving the environment," something that I learned to respond to with, "I'm more interested in how racism affects the environment." What I didn't realize until much later was that when Jane said "walk the trail," she wasn't pointing me to a nature trail—though it was that, too—she was pointing me to a place from which to orient myself, a different starting point (in both space and time) from which to understand toxic capture. This place, where wasteland and wetland slide back and forth in the *longue durée* of Black un/freedom, is a place where the histories of enslavement and settlement continue to rob life in and from land. Despite the way colonial historiography forecloses nonpropertied stories about environments, this is also a place with many names: the Great Dismal Swamp; the homelands of the Nansemond and the Algonquian speaking tribes of the Tsenacomoco alliance; and the Underground Railroad.

But we'll get back to that.

In the parking lot, the girls had made living room: something working girls, chronically condemned to expulsion, make in unexpected places, and Jane had walked me over, invited me in. We sat down and made ourselves comfortable on the plastic chairs scavenged from the dumpster, abandoned tires found on the side of the road, and boxes that had been discarded by the 7-Eleven. The parking lot had been transformed into a kind of refuge, where their experiments with waste's matter and meaning made places to tell stories about their own surplusification and contemplate their un/freedom.

"Let's tell baby girl here about Suffolk; don't you think she should go?" Jane queried.

"Oh lord, yes," Mary affirmed. "In Suffolk, racism is R-E-A-L," she said, drawing out every letter.

"You can see it everywhere," Betty added, nodding her head.

Mary, this time: "It's like ain't no time has passed."

Almost clipping the end of Mary's sentence, Jane piped in: "Ain't no time *passing*, Mary. You can't leave. The foundations are so bad, so you can't even sell your house. You can't sell your house; you can't go nowhere. People work close to home, and all the brothas have the same job at the peanut factory if they got a job at all. Nothing has changed. You've gotta know you're in the south now. The n****s out there be slow as molasses, like . . ."

Jane stood up and, like she was in swamp water, mimed an exaggerated walk in slow motion. There was something about how she moved that I would come to realize, was more than a way to comment on a kind of Southern masculinity; it was diagnosing something about the way history *pools* here, a way time slows or the past endures in the present. Perhaps her movements intimated a science-fiction-like feeling where in Suffolk, the past feels too close, like you could stick your arm through a wall and get sucked into a Southern plantation.[2]

This chapter takes you on my time-jumping journey across the repetitive terraforming projects that make Suffolk a place where settler-colonial land use requires the mattering of Black wageless life. Shifting from Marx's titular revolutionary subject—"the waged worker"—I center the perspective of wageless Black women to reveal how white supremacy is a terraforming project that *requires* the continued use of Black flesh (not just Black people as waged laborers). The analytic focus on the waged worker, at times, obscures how the commodification of land for capitalism demands ongoing settler-colonial violence that relegates indigenous sovereignty and the wageless slave to the dustbin of history.[3] A focus on waged labor, and waged waste-labor in particular, can make it difficult to see how making life surplus to land (through property's violent sociomaterial relations) plays an ongoing role in settler-colonial racial capitalism. However, it would behoove us to remember that, as W. E. B. Du Bois argues in *Black Reconstruction*, the slave was a *wageless* worker.[4] Under chattel slavery, Black flesh was rendered a *tool* to transform land in the interests of the white settler-planter, but the slave's *personhood* and *presence* was rendered an excess to (stolen) land.[5] Wagelessness, I contend, is a relationship to land, a relationship that continuously dispossessed and condemned girls meditate on, a relationship that, as I will define below, is what I call becoming fill.

From the monocultural violence of tobacco plantations and the attempt to drain the Great Dismal Swamp in the 1700s, to the settler frontier of present-day landfills and the devastated soils under Black homes that are worth *less*, this chapter moves back and forth across the political economies that have transformed Black life in this settler colony. This movement reveals how surplusification, land devastation, and toxification make Suffolk a place of pooling environmental injury, including enslavement's *longue durée* of the injury of racialization. This injury turns Black life into the matter (fill) that stabilizes settler property regimes, making use out of the destruction that facilitates white life.

Between the landfill and the swamp, where the high risk of poisoning maps onto Black residences, the (de)valuation and surplusification of life and land is a living history written in the materiality of discarded elsewhere(s).[6] And those materials include waste, people's bodies, and the more-than-human flesh of our biosphere. As we saw in chapter 1, the threshold theory of pollution turns Blackened life into a sink for acceptable risk, and those risks accrue. But in order to understand how toxic capture emerges from the *longue durée* of anti-Blackness as colonialism's environmental management, I want to point us to the logic by which Black people become surplus to and for land. If toxic capture marks *use* of Black bodies (not simply as workers or laborers), the logic by which Black people are cast outside nature requires the surplusification of Blackness to the environment.

TOXIFICATION AS SURPLUSIFICATION

Toxification is a form of surplusification that reinforces the way that "the production of surplus value . . . is a shrinking form of wealth extraction."[7] Labor exploitation, which Karl Marx so famously diagnosed as a system that robs the worker and the soil, "is transitioning (always unevenly) to outright and violent *expulsion* and disposability."[8] And as Neferti Tadiar argues, the production of value has always been part of the colonial *violence of humanization.*[9] "The war to be human . . . [is] the war on the part of the already human to remain so . . . with all their proxies in tow," while "becoming-human in a time of war . . . [is] the struggle to live on the part of those who can never be fully 'human,' the abiding category for valued existence or what the global deems [a] life worth living."[10] Humanization, and the project to secure the human's categorical boundaries, are thus nefariously linked to the violence of making human life a *value-constitutive* activity for capitalism.[11]

The process by which Black people were "granted" their humanity was the process of becoming waged workers, prioritizing waged work in our Black political and perhaps environmental imaginations.[12] However, it was also precisely along the raced and gendered lines of waged work that the slave's *conditional* humanity was reinforced, recasting the expropriated African as a now "free" and exploitable American labor force. Becoming human, so to speak, was predicated on what Robert Bullard once called job blackmail: "Black workers are especially vulnerable to job blackmail because of high unemployment and their concentration in low-paying (high-risk)" jobs.[13] The wage didn't so much grant Black people full humanity (much less self-determination), as much as it became a vehicle through which to enforce white supremacy's patriarchal geographies of racial violence. Jim Crow reinforced white divisions of gendered labor, narrowing where Black men and women could work. Moreover, separating Blacks from whites translated the color line into types of labor (domestic and manual, "dirty" and "clean").[14] Thus, segregation was a useful way to produce hyperexploited Black workers through spatial (and environmental) condemnation.[15] To rephrase Robert Bullard's point, job blackmail is always environmental.[16]

These examples abound. In Pavithra Vasudevan's analysis of the present-day alchemy of race, intimate inventories of Black waste workers reveal how Alcoa (Aluminum Company of America) outsourced toxicity to the Black families of segregated Badin, North Carolina. With the help of one of her interlocutors, Raymond, who worked at the smelting factory, she notes how those "who endured industrialization's slow violence are marked by toxicity, a material manifestation of atmospheric racism."[17] J. T. Roane's work on the Black ecologies of subaquatic life in the Tidewater traces how the industrialization of the Chesapeake Bay sits in stark contrast to the agriculturalist ethics of Black oystermen in the 1880s and 1890s who sought "stability and basic social and familial soundness" to sustain their small holdings, "not necessarily massive wealth derived from an ever-expanding holding."[18] These practices, which protected the regeneration of oyster populations and forwarded a vision of "interdependence between the land as well as salt and fresh waters," were quickly put at odds with a system of codification that required licensing, taxation, and fees for leasing the river bottom.[19] What's more, strip mining and mountaintop removal in the Appalachian Mountains had devastating ecological effects. Poor white and Black people subject to the brutalities of Jim Crow were reminded that segregation was part of a labor regime in which the health consequences of extraction materialized in the body as much as the land was scarred and depleted—riverbeds filled in with excess rock, and waterways choked by run-off sediment.[20]

The violence of becoming human is of particular importance to tracing the material continuities between chattel slavery and its toxic afterlives. If producing value is part of what defines one's relationship to the human, not only does white supremacy's land-use politics punish people of color through inhumane environmental contexts of their labor, but the toxification and destruction of delicate ecologies is the condition for producing surplus people who cannot attain "full humanity" through "the wage."[21] The toxifying effects of capitalism's regimes of racialized extraction render Black life-making as not just quintessentially surplus to *value* but quintessentially surplus to the *land*. Thus, while an analysis of *labor* and its environmental politics are critical to retheorizing the environment, this chapter argues that we need to see the status of wagelessness as fundamental to "colonial looting, slavery, . . . imperial wars central . . . to world capital," and the specific land/ecological politics of settler colonialism and racial capitalism in the United States.[22] As outlined by Eve Tuck and K. Wayne Yang, the "settler-native-slave" triad upholds white possession as the science of land use.[23] For the settler, Indigenous peoples are "in the way" and must be destroyed in order to recast "land . . . as property and as a resource." On the other hand, "the slave's very presence on the land is already an excess that must be dis-located."[24]

Racialization functions to mark "the slave" with the status of being *excess* to land, or what I call "unecologic." This eco-logic of chattel slavery inaugurated the contradictory status of what Charisse Burden-Stelly calls "value minus worth," the quintessential disposability of Black life.[25] The attendant "series of *surplus geographies*" created in "a world shaped by money but populated by the moneyless" has a long history and takes many forms across the globe.[26] It is critical to the relentlessness of privatization's enclosures meted out by the geographies of accumulation and dispossession that recursively come to define "a life worth living." If the worker, as Marx so famously analyzed, *always* transforms nature through his labor power, Marx underestimated humans' relations to nature, including for nature to transform labor. Settler colonialism's usurpation of Indigenous homelands, and racial capitalism's disposability of Black flesh, are in a nefarious relationship to the figure of the worker. Alongside the worker, whose labor value (even if low waged) is definitive of the exchange of a wage for the freedom to buy a life worth living, capitalism has simultaneously constructed a wageless surplus from which it steals and transmogrifies life and land.

In this chapter, I want to draw out the eco-logic (or "*un*ecologic") that makes Black people complexly (even if contradictorily) linked to land. The quintessential disposability of Blackness is a form of surplusification that materializes through a process of "becoming fill." In the language of development, construction, and engineering, *fill* refers to materials that "fill in" or stabilize the foundation, creating the ground on which to build. The logic by which Blackness is cast outside the environment (turned into matter to be used to make land pollutable) is also a way of making Blackness "unnatural," "unecologic," an excess to land. The unecologic is a way to make Blackness available for a land management strategy in which Blackness is always matter that matters for development. Ironically, as surplus matter, Blackness is turned into a material excess to land that must be managed (waste) or put to environmental use (fill). Thus, becoming fill bears the weight of contradictory explanations and processes of anti-Black capital accumulation. Under racial capitalism, Burden-Stelly argues, "Blackness is a capacious category of surplus value extraction [and simultaneously] the quintessential condition of disposability, expendability, and devalorization."[27] For the purposes of *becoming fill*, I take up this dynamic to reveal how fill is always made excess to land and thus put to use *for it*.

Before I continue, I want to emphasize that while my thinking on surplusification might be applicable in other places, I also want to be specific about the work Blackness is made to do in the United States. After all, if this argument was made to so easily transplant across the world to where every Black life lives, it would subject Black *people* to the condition of fungibility. This condition might describe the anti-Black logics of value production, but it does not emanate from Black life in its multiple forms, nor does it describe "the labor of living" in its geopolitical and historical specificity.[28] Also, I see surplusification as an inherited condition of enslavement and the "settler-native-slave" triad foundational to the eco-logics of property in the United States. Within the constraints of this triad, racialization consigns Black and Native to different but intertwined statuses of dispossession in order to produce white supremacy to which all minorities are subject—including in their hyphenated and mixed-race identities that are far more plural than the structure of white supremacy's triangulation can name.[29] But because of the way Indigenous removal and the slave-as-excess structures property and the science of land use, this eco-logic remains an important environmental operation to unpack. This triangulation of racialization is foundationally injurious, including in how it consigns *all* nonwhite people to competing and often incommensurate statuses, and my hope is that the figuration—or, rather, disfiguration—of Black as "excess" or unecologic might change the intersections of who, what, where, and when we learn from in the spirit of abolition ecologies.

Fill is thus a way to see how Blackness is matter(ed) through and for land, a way to notice Blackness as an ecological relation inaugurated by chattel slavery in this settler colony, and to see racism and racialization as a project that always terraforms environments. If becoming fill is a way that Black lives come to matter for land, this chapter seeks to understand stories of surplus, where the wageless, not the worker, can tell us how, to put Marx's words differently, capitalism simultaneously robs Black life and Indigenous land. Inspired by Vanessa Agard-Jones's insurgent reading of the colonial histories that the sands of Martinique remember, noticing fill is a way to notice the liveliness of what David Silkenat calls "scars on the land."[30]

"Our houses are sinking," Mary said of her family home in Suffolk, Virginia. The town is shaped by the past monoculture of plantation economies as much as by the surplusification of Black life. And these economies play themselves out through the alteration of Suffolk's soil. For wageless Black women displaced by "bad" soil, sex work transformed the racial-sexual economy of geographic condemnation into a way to move. And "moving as I please," as Jane said of her relationship to sex work, was "a way to fuck with [a world where] everything [was] always trying to pin me down." But the soils to which these women point are moving, too. In Virginia, the place that inaugurated the eco-logic of property whereby "Black" is always surplus to and for land, their attention to the soil tells us about the material conditions of turning Black life into surplus matter.

"It was like a funhouse growing up there. Like, every day the ground would move and the house would change. Not all at once you know, but by the time I was maybe 6, I could put my arm through the place where the wall was supposed to meet the floor," Mary said. Throwing the box of cigarettes to Jane, Mary passed the proverbial mic.

Jane lit another cigarette and shared, "I remember in the front room of our house, near the ceiling, there were cracks in the wall. Girl, they would just get bigger every time it rained. My mama used to get so mad at me and she would be yelling and pointing to the wall and say, 'You're gunna tear the damn house down.'" The circle broke out into uproarious laughter and affectionate exchanges. Turning condemnation into congregation, laughing was a way to say *I know what you mean*.

The fine loam soils of Suffolk have been radically altered and are part of why the homes lean. The soil is sandy, making it prone to erosion, but it is also porous, making it prone to flood. Ironically, it was the sandy soils of Suffolk that brough English colonists to Tidewater, Virginia. The geomorphic qualities of sandy loam (light and well-draining) make it a soil in which tobacco thrives. However, centuries of soil exhaustion, as well as colonial interventions into the landscape, have affected the soil's integrity. This once well-draining soil, now home to residential development, is all the more fragile. As the biosphere warms and the atmosphere holds more and more moisture, the soils flood.

Tobacco planting dominated Tidewater, Virginia, agriculture from the early seventeenth century to the period of the American Revolution. Like all plants, tobacco requires specific environmental conditions. In soils that are prone to poor drainage, tobacco roots will die from lack of oxygen.[31] But tobacco is thirsty and "needs an adequate amount of water to maintain turgidity and expansion of the leaf."[32] Colonists, however, did not plant tobacco in concert with other things, which would've been more in line with the agricultural lessons learned in England. Instead, colonists used enslaved labor for massive land-clearing projects, and monoculture became the agricultural method by which the planter class would amass its wealth.

Unsurprisingly, the soils were quickly shorn of their nutrients, and just as quickly as the land had been cleared, planters moved westward, expanding the plantations' reach. By the early 1800s, planters had abandoned the fields that once stood "at the epicenter of Virginia's tobacco boom," marking the ecological devastation that produced "America's first planter class."[33] Environmental historian David Silkenat argues that soil erosion was critical to the decline of tobacco cultivation in 1780s Virginia: "Southern plantation agriculture began on a fragile ecosystem" and its voracious devouring "leached the soil of nutrients and promoted erosion."[34] Ecological devastation, including the alteration of soils, was in no small part "fueled by slavery's expansion" and the voracious search for fertile soil. "No environment could long survive intensive slave labor. The scars manifested themselves in different ways, but the land too fell victim to the slave owner's lash."[35]

Brian Williams and Jayson Maurice Porter argue that plantation monocultures "build [ecological crises] over time."[36] While their argument is dedicated to cotton's specificities, all "plantations are nutrient-hungry, deriving profit from the exhaustion of land and soil." As much as the planter class could bury these ecological liabilities in the future through the constant expansion of the colonial frontier, Williams and Porter argue that these consequences would eventually lead to a series of chemical "fixes" for the ecological crises in the form of synthetic pesticides and fertilizers. In other words, the eco-logics of the plantation would lead to ever-evolving forms of landed subjection and dominion.

The accumulation of harm in the Tidewater, including harm to Black life, is the product of histories' future crises buried in the land and in the soil. "Not just my house, a lotta people's houses," Mary said to us in the parking lot. "You just go drive around, girl, you can see [the soil] makes our homes lean. I knew a girl who came through here once said her house just fall down, like it caved in on itself." As Jane said, "Our houses ain't worth *shit*," or as Betty asserted, "Ain't no value in swamp trash." Both home and body become surplus matter through the valuation of property. And this production of surplus matter will be put to use to stabilize the ground(s) of white futures.

"We're still stuck," Betty cried, "I ain't gunna be stuck no more!" When I finally asked what they thought it was about the soil that caused their homes to sink, Betty laughed and said, "'Cuz it's trash," to which Mary added, "Because it's haunted!"

The plantation haunts the ground and it's wasting haunts the present and the future. As (part of) the blueprint for a settler-planter economy, the plantation was also a terraforming project. The monoculture of extraction, the commodification of land, Indigenous genocide, and the lash of enslavement left a legacy in the soil. While Virginia's planter class created the conditions for their lives to be the ones "worth" living, they did so by wasting futures, or rather, by treating the conditions for future life—the complexity of ecology—as disposable. Producing the conditions for the surplusification of nonpropertied (and more-than-human) life forecloses the livability of other socio-ecological arrangements. Though I can only speculate about what role soil plays in making Black homes "trash," the soil might nonetheless be, as Agard-Jones argues, an important "textual archive."[37]

Moreover, when Betty says the soils are "trash," it is not just metaphorical; trash is what stabilizes the post-plantation grounds of development. Referring to the uneven development of Suffolk and its worth *less* sinking houses Betty says, "You *know* that shit is built on some *swamp trash!!!*" Bursts of laughter ripple through the circle as the girls continue to speculate about their relationship to land.

"Well, and those new companies!" someone adds, tracking the construction and development transforming the Tidewater into the Hampton Roads.

"What's that one we're always seeing at night now?" Janes asks. "Ryan Homes," Betty says, as if she has said it many times before, like a reminder: Pay attention. "We see trucks late at night, all the time now, just dumping rubble into weird places and shit. I saw them the other night [while hookin',] dumping trash into the swamp. Plain. As. Day."

In 2019, rumors of secret settlements with homeowners in South Carolina began to emerge in the local newspapers, on personal blogs, and on Reddit threads in the South.[38] The accusations against Ryan Homes were varied—shoddy construction, unconnected toilets, uneven foundations—but a common thread emerged over and over: Housing complexes were being built on unstable foundations of waste. Coincidentally, Ryan Homes hired Waste Management Inc. (WMI) to manage its construction debris while building in the Tidewater. As a waste conglomerate setting up a corridor of waste flow from North to South (see "Flow"), it is not a surprise that WMI was also managing construction waste, the surplus matter of development.

The women at the motel constantly speculated about where things were being dumped: their backyards, their childhood homes, behind the projects, and in the swamp. "Well, we all know that we've been poisoned. Ain't shit we can do about that now. Probably not even then! You know these white motherfuckin' Bill Gateses building homes don't know shit building all over this swampland. They ain't never did." Though I never saw the conspicuous dumping, I did see Ryan Homes constructions all over the Tidewater. That the girls regularly tracked the development company's presence with mistrust suggests something routine about property enclosures, the girls' impending expulsion as surplus matter, and their consignment to the environments of toxic capture. As Tuck and Yang argue, in "the settler imagination," the slave "is reconfigured/disfigured as the threat."[39] The slow death of toxic capture is the materialization of this configuring, and the inheritance of being a threat to property makes the possessive grammars of land use ecologically *unsound*.

Like Mary said to the chorus of nods and *mm-hmms*, "It's a real problem." Her father's house had been built on "shrink-swell" soil, the "fine sandy loam" that Virginia's planter's class had once revered and just as quickly abandoned.[40] His home is *literally* sinking into the ground. "He prays for it not to rain." We all nodded, grunting noises of affirmation. Mary's story is not unlike others that have been covered in the news: "He digs a trench around his house hoping that the rains won't flood the house or that the toilets won't overflow. But mostly, he hopes that heavy rainwater won't bring waste."[41] In 2021, the *Guardian* reported on a wastewater spill that "brought to the surface the dangerous impact of the region's" infrastructural racism. Many Black homes are disconnected from the region's sanitation system because the soil is unsuitable for septic tanks, and the pre-existing sanitation system is aging, making it prone to overflow from the pressure of heavy rains. Compounded by drainage problems, Black homes, like the sinking homes that began this section, become vulnerable to sinking, flooding from stormwater, and the presence of wastewater overflow from a sanitation system that they don't benefit from. Moreover, the stormwater, which "flows in[to] open ditches and is released into larger above-ground culverts and retention ponds that make up the city's drainage system . . .

can pick up septic tank runoff, fertilizers, sediment, pet waste, silt and other pollutants that can contaminate waterways and damage coastal habitats."[42] As if that wasn't enough, the city's "stormwater runoff feeds down into the culvert networks of Pughsville, a historical African American village on the city's border with Chesapeake that has overwhelming drainage problems of its own."[43]

What would it mean to hear these women diagnosing the processes through which Black life is turned into surplus matter? How might their attention to soil orient us toward how Black life is rendered "unecologic," surplus to the environment, and yet a threshold where harm to flesh—Black and biospheric—accrues? If soil is an "archive" through which enslavement's ecological crises endure, perhaps seeing scars on the land is a way to notice that the mattering of Black life is already shaping the future? Will Blackened Suffolk become a modern terra nullius that white property regimes can seize? And if so, what will land and life be seized *for*? The stories of Black surplusification (through toxification, sinking houses, or becoming a dumping site for construction waste) are stories of fill in the making. Past futures and present crises always seek a fix for surplus, so what use does making Black life fill have?

MOTIVES BEYOND MONEY

While soil was a way for the girls to name how racist histories are made material, devaluation makes the soil a site of diffuse racial violence.[44] According to the first soil survey of Suffolk done in 1981, much of Suffolk is not fit for development. What is at issue is the "marshiness" that comes from the "shrink-swell" soil.[45] The soils of Suffolk are of "very limited" use for development, especially residential development.[46] As the *Guardian* noted, "90 percent of Suffolk is not suited for septic tank systems because the soil is too sandy."[47] The strategies proposed for residences built on these soils is to raise them, a capital-intensive project that draws an economic and ecological line around the poor, of which Suffolk and the historical African American village in Pughsville, near Chesapeake, are both examples. But the soils are not only unsuited for septic tanks; they are also not suited for sewage lagoons, sanitary trenches, or landfills. While undoubtedly "soil science established a technoscientific grammar for the chemical consolidation and expansion of colonial capitalist farming," even the capitalist profit-derived classifications that "flatten and fix soil" properties in the survey clearly suggest that Tidewater soils are severely prone to seepage and wetness.[48] This is a critical factor in the placement of waste infrastructure.[49] Still, Suffolk is home to the Southeastern Public Service Authority (SPSA) landfill, a regional public landfill.

It might be surprising that on the other side of Black Suffolk is the largest known swamp in the United States: the Great Dismal Swamp. During the conservation movement that coalesced in the 1970s, conservation shifted from natural resource management to wildlife habitat protection, articulating the importance of wetlands to biodiversity. In 1974, the swamp was designated a wildlife refuge, which protects the swamp as an important wildlife habitat under the US Fish and Wildlife Service. While swampy ecologies are critical to wildlife, swamps are also critical to waterways, marshlands, flood plains, soil health, and coastlines. I point this out because the Great Dismal Swamp and its multiple ecological relations have been altered by a great number of colonial projects, including Route 58 (the longest road in Virginia), which alters the swamp's relations with the surrounding wetlands.[50] As the colloquial name Tidewater might suggest, the land is watery and the more-than-human life in which land is entangled is subject to the delicate balance of the ocean's tides. Both the road and the wildlife designation don't so much protect a habitat as they impoverish the environmental imagination of the complex and fragile connections that development alters. Route 58 divides the swamp from the wetlands north of the road and is the cartographic boundary with which the landfill imagines that it is far from protected wetlands.

Route 58 is also how you get to the regional landfill.

Since its inception in 1976, SPSA has been a contentious regional co-operative. Taking out public bonds to finance a regional waste infrastructure—landfill construction, waste transfer stations, an energy plant, and a leachate processing site—SPSA came into being in a cloud of debt. To make matters worse, the tightening of landfill regulations by the Environmental Protection Agency in the 1990s caused a cascade of landfill closures at a time when city services were privatizing. The cost of retrofitting closed dumps threatened to bankrupt a number of coastal cities, and private waste contracts offered a cheaper solution. For public cooperatives like SPSA, however, this posed a crisis: If giant private companies could offer cheaper and more competitive contracts, SPSA would be driven out of business. Struggling to stay afloat in a quickly privatizing market, SPSA took on even more debt to keep up with the low tipping fees that private companies like WMI could provide.[51] For public landfills, whose contracts are often smaller and limited by the capacity of the landfill itself, tipping fees are pegged to the lifespan of the landfill. With the regional landfill set to reach capacity by 2012, the viability of SPSA hinged on expanding its landfill and, perhaps most importantly, on its ability to generate capital to fund the expansion.

Private waste contracts with cities like New York and the rapid construction of seven private mega landfills in Virginia in 1993 made for massive flows of waste into Virginia. Without the volume or disposal capacity of a giant private company, the value of SPSA's disposal capacity depreciated quickly, making it increasingly difficult to pay off debt with the revenue generated from tipping fees alone. In order to generate capital, SPSA needed more waste, and to stay afloat the company would have to find a way to control waste flow. In the waste industry, "flow control" is not a metaphor; it is "a legal provision that allows the state and local governments to designate where solid waste (MSW) is taken for processing, treatment, or disposal." In this new market, those who direct flow control direct the flow of profits. And importantly, "Because of flow controls, designated facilities may hold monopolies on local MSW and/or recoverable materials."[52]

As a first attempt at controlling waste flow, SPSA tried to invoke a flow control statute to get the state to act. SPSA wanted to hold a monopoly over waste in the region. But as the Supreme Court would decide in 1994, state control over the flow of waste is at odds with the "spirit" of capitalist markets. When the Town of Clarkstown, New York, attempted to sue C & A Carbone Inc., a private recycling company, for shipping nonrecyclables across state lines, the Supreme Court decided that the town's flow control ordinance violated the Commerce Clause and was unconstitutional. The Supreme Court argued that the ordinance "discriminates, for it allows only the favored operator to process waste that is within the town's limits . . . squelch[ing] competition [and] leaving no room for outside investment."[53] While flow controls are critical to the jurisdiction of disposal sites the C & A Carbone, Inc. v. Clarkstown decision set a legal precedent for private waste-processing companies to exploit flow controls to secure their own contracts. By favoring privatization, the decision shifted the ability to monopolize to waste conglomerates, rearticulating waste management as a market industry and not, say, a public good.

Unable to compete with the capacities that private waste companies could provide, SPSA engaged in a series of debt mitigation strategies. First, it began cutting costs—including facility maintenance, leachate monitoring systems, and the maintenance of boilers at its waste-to-energy (WTE) plant. Second, it sold its WTE plant to one of WMI's subsidiaries, Wheela-brator, which now owns and operates the hazardous site choking Black life in Portsmouth (see chapter 1). Third, it began trying to partner with large waste corporations, hoping that investing in building giant waste ports and mega landfills would pay off in the future (see chapter 1). Finally, it began taking in more waste than its landfill could handle, attempting to use the extra money to offset the cost of its more capital-intensive strate-gies. When SPSA was eventually audited in 2008, a local reporter from the *Virginian-Pilot* summed it up succinctly: "At bottom, SPSA's current woes are the inevitable result of burying liabilities in the future to avoid paying them in the present"—not unlike the tobacco plantations that characteristically ravaged the land.[54]

While the reporter might have been using "burying liabilities" as a metaphor for fiscal mismanagement, it's an acute material reality of racial capitalism. Landfills are liabilities buried in the future, and the future is buried in the ground. And as landfill engineers know, landfills always leak; it's not a matter of *if* but *when*. SPSA's risky debt mitigation strategies—especially its neglect of maintenance and leachate monitoring—would lead to a massive leachate buildup in 2018. However, years before the thirty million gallons of leachate buildup made industry news, state monitoring of groundwater suggested that leachate was seeping into the ground after the landfill began to spring surface leaks in 2003.[55] Traces of lead, cobalt, and beryllium were found in the wetlands nearby, but SPSA refuted that a landfill leak was at fault, arguing that those same elements are also found in swampy soil.[56]

In "Chemical Regimes of Living," historian of science M. Murphy ar-gues that our collective chemical relations are "the accumulated result of some two hundred years of industrialized production" and the result of externalized effects "posited as existing outside the accountability of corporations, and in the context of neoliberal governments, outside the scope of regulation."[57] Like the presence of lead, cobalt, and beryllium here, toxic relations are externalized to the swamp, mystifying the direc-tion or cause of toxic capture. As I argued in chapter 1, it is precisely the multidirectionality of ecological violence that makes toxic capture mun-dane and unprovable. But more than distributing or outright avoiding ac-countability, SPSA's behavior points to capitalism's surplusification. From the lives poisoned by living in high-risk Black zip codes where homes sink and flood, to the many organisms and histories connected to the wetlands, being turned into surplus life is always material, "rooted in the flesh of the planet."[58] Making toxification external to the landfill but internal to the wetlands ecologically condemns all those who are connected to it.

It seems an opportune coincidence that the year before lead was found, SPSA purchased 525 acres of undeveloped land adjacent to the currently operating landfill. As previously mentioned, one of the company's debt mitigation strategies had been taking in more waste than it could handle. Before the city of Suffolk had even approved the land-use permit, residents noticed that SPSA had already begun clearing the land for new cell construction.[59] Landfills are divided into parcels of land called "cells." Each cell is equipped with a special liner to catch moisture in order to slow the seepage of leachate into the ground. Waste is then dumped into a cell and covered with a layer of soil to deter pests and smother odors—ironically, soil cover is one of the few things that Suffolk's soil survey deems sandy loam good for—and this process is repeated until each cell is full. With cells 1–5 already full, and the final cell, cell 6, in operation, SPSA had already begun planning new construction for cell 7. In the wake of the 2008 audit, SPSA was more determined than ever to pursue a landfill expansion plan and tried to fast-track cell 7 without the town's permission.[60] Immediately, the expanding frontier hit a regulation boundary. But it was not the quality of the soil, as the 1981 soil survey would suggest, and it was not the presence of lead thought to be coming from SPSA's landfill; instead, the construction of cell 7 was three-quarters of a mile too close to a human water supply.

Local reporting on the stalled cell 7 expansion plan, however, reveals confusion about the location of the water supply. In yet another string of articles chastising SPSA's failures, this time the focus was on spatial mismanagement rather than fiscal mismanagement. At first, the local papers reported that the groundwater well in question would affect the residents in Green Pines Apartments. However, these same early articles noted that "the name and location" of the well in SPSA's Department of Environmental Quality (DEQ) application "could not be verified."[61] Wilson Pines, an apartment complex owned and managed by a private company, is one of the few complexes in Suffolk that accepts Section 8 vouchers through the Department of Housing and Urban Development, and might've been the misnamed complex on the DEQ application. Most of the Wilson Pine residents are Black, all of the residents are low income, and the complex is located in one of Suffolk's two high-risk zip codes, next to the Great Dismal Swamp and across Route 58 from the landfill. Another article, still uncertain about the name of the site, made it painfully clear that expansion was the priority, no matter the cost: "Perhaps the best solution for SPSA is the simplest.

For the millions of dollars that are tied up in the landfill expansion, SPSA should buy the intruding apartment complex, relocate the residents and demolish the site."[62] Be it the callous disregard in the reporters' words or the disregard manifest on the SPSA application, the poor people proximate to the landfill are matter easily displaced. "If we only understand surplus life as an economic question of employment or unemployment," we miss the way that "surplus life pivots on a deep geographic injustice."[63] According to this reporter, it is not only more profitable but *logical* to expand a landfill at the expense of disposable life. The confusion and misnaming of the "intruding complex" signals an "excess" to be removed in the spirit of market arrangements and colonial land use. The landscape of environmental regulation and waste infrastructures rely on the production of and subsequent usability of surplus geographies. And the disposability of Black life is quintessential to the manifest seizure of land.

The solution to SPSA's cell 7 water well problem came just months later.[64] A letter issued by the DEQ stated that the water supply for the still-unnamed apartment complex was elsewhere. In his response to SPSA pleas for a work-around—again, in 2008, SPSA's viability hinged on the expansion plan—then-DEQ Executive Director David K. Paylor granted a conditional-use permit to build cell 7, stating that "the expansion site is downstream of the water supply in question, at the Green Pines Motel East, and that the [well-water] rule is concerned only with landfills that are *upstream* of drinking water sources."[65] With the complex finally named and a work-around acquired, SPSA was well on its way to servicing the "public," and recasting it's expansion, and toxification of low-income Black residents, as a "good."

Environmental regulations and surveys politicize the topographical direction of landscapes, constructing causal relationships that are not necessarily accurate. While the regulation stipulates that landfills must be *downstream* of a water well, if the landfill is abutting a water well but it is *upstream*, then the regulation has *not* been violated. In other words, when push comes to shove, landfills can be brought closer and closer to water wells (as long as they are not serving white residents, of course) when expansion is deemed necessary. Every possible place where the well might've been located was Black and poor. Whether the water well was near a motel or sinking houses, no one bothered to *place* the risk, because Blackness is an acceptable threshold for polluting land and the disposal capacity for the nation.

Ten years later, in 2018, when SPSA's contract was renewed and the company was legislatively required to appoint a brand-new board of directors, each board member—a representative designated to advocate for the economic health of their municipality—was chosen for their financial expertise. Suffolk's board appointee, however, brought a different expertise to the table. Suffolk appointed an eminent domain lawyer with a history of successfully defending polluting industries against environmental lawsuits. For a lawyer with a long track record of successfully defending corporations' constitutional right to pollute, the disposability of Black life is a good way to toxically entangle soil with money.[66]

Like all frontiers, the imagined boundaries of the landfill are factitious; they are *impositions* that parcel land for use, including as a public "good." "Land as a Resource, a practice that . . . generate[s] pollution through pipelines, landfills, and recycling plants, or as sinks to store and process waste," is a logic of colonialism. Or, as Fanon might articulate, the colonial lines of violence are upheld by the butt of a gun. In this case, the gun is the "surplusification of life" and its attendant "surplus geographies."[67] Back in the circle, Jane says, "My mama's house is just falling apart because we's Black and we live in the South. They build our houses on some trash, and then they let them fall down and they tell us it's our fault? How does that even make sense?"

"It's true, Mary says. "But it's like, I don't even care. I don't want to stay there because it's like everyone is stuck."

Betty picks up where Mary and Jane left off: "It's not like anyone cares about our homes; they will disappear eventually, and we will *not even be* history. My uncle is a good man, worked all his life at that factory but ain't nothing come to him as a result of it. He just worked and worked and worked and worked. And now he's old and can't work no more. It's like Mary's said, the ground gonna take him."

The Blackening of place is always a way to provide an arrangement for risks to accrue, and these risks turn Black life into a material excess.[68] And, it should be noted, these risks accrue harm to more than human life. Betty viscerally articulates how her uncle is becoming fill: "The ground gonna take him" is an all-too-eerie articulation resonant with Katherine McKittrick's description of the African burial ground discovered beneath the New York Stock Exchange: "The . . . ground tells us that the legacy of slavery and the labor of the unfree both shape and *are a part of the environment we presently* inhabit."[69] Now wageless and stuck, Betty's uncle is becoming fill for a future environment.[70]

Becoming fill indexes something unsettling: The foundation is not, as settler colonialism would have it, quite so *settled*. "Sometimes it don't drain, sometimes it's too dry. It moves or something," Mary notes of the sandy soils beneath her father's home, reminding us that foundations are shifty. Wet land as land for waste implies that things move—water carries lead, leachate viscosity moves toward the water table—and teach us that property is an imposition of relations onto the land. Becoming fill is an analytic that asks us to contend differently with the force of naturalizing toxic capture. It is also a way to insist that though disavowed by the mattering of Black flesh for land, Blackness is an ecological relation and wagelessness a relationship to land. Surplusification is the prime condition for toxification, and making fill is racial capitalism's attempt to naturalize its own processes of becoming, especially because things, as Mary suggests, resist settlement.

The night before I headed out to Suffolk for the first time, Mary warned me in private: "You be careful where you step, girl. Things are shifty out here. You think you know something and you'll quickly realize it ain't true." I thought she was talking about men, warning me as the other girls had about getting caught up with someone who might tie me down. But then Mary said, "Don't ever think you're standing on firm ground. Out here, things are never quite what they seem. Don't get stuck, you just gotta keep moving." In what felt like a maternal warning about the demands on Black women, a warning about how patriarchy keeps women's movements confined, I also heard an ecological ethics that emanates from dispossession. The environmental knowledge required outside a propertied "life worth living" is to be "among [one's] own in dispossession, the ones who cannot own, the ones who have nothing and who, in having nothing, have everything."[71] Mary's warning about the ground is also knowledge about racial capitalism and settler colonialism's relationship to it, wherein a warning against settling is a warning against the promise of settlement. Property is a false promise that these women know well, and ownership for those so easily turned into a surplus is not a way out of the condition of "value minus worth."[72] Mary warns, "Don't let nothing suck you in," including white supremacy's possessive eco-logics that make you (into) matter to commodify the earth.

TERRA FIRMA

Racialization is an ongoing form of environmental injury. With respect to Blackness, the eco-logic by which this manifests itself over and over again casts "Black" as excess to property and "unecologic." In Virginia, where the present conditions of possibility for Black life make race, waste, and the swamp proximate, the histories of colonialization's injuries *pool*. Wetland and wasteland slide back and forth in the *longue durée* of Black and Indigenous un/freedom. And Blackness is an ecological relation to land forged by the punishing science of plantation land use extended by landfill expansion. But it is the surveying of land for the violent parceling of plantations to which I now turn. The women at the motel never made direct reference to the plantation, but statements like, "We're *still* stuck, you know," and, "Nothing has changed. You've gotta know you're in the south now," resonate with what Hartman has called the afterlives of slavery: "skewed life chances, limited access to health and education, premature death, incarceration, and impoverishment."[73] In Suffolk, where Black homes are sites of accumulated risk—including sinking and poisoning from the ground or dispossession for waste disposal—the houses are trapped by a material economic order that materializes the settler-planter antebellum belief that the wetness of land was equated with "a waste of it."

Forged in the violent removal of Indigenous people from the land, enslaved labor was used to terraform colonial "wastelands" into infrastructures for white life. These terraforming projects were ecologically devastating, including through the use of enslaved labor as a tool of ecological destruction, rendering the slave into an ecological weapon. The tobacco planters that altered the sandy loam until the 1780s were equally as invested in "improving" swampland. As told by those who authored settlement, cleaning up Indigenous "wastelands" was a moral duty. Absent from their "heroic" stories, however, is the story of Black being/ecology. As both the tool of ecological transformation and persons whose presence was considered an "excess" to the land, slaves, I argue, were not considered ecological at all.

Slavery was an environmental project, and the environment was a colonial matter of concern.[74] Constantly in search of rich fertile soil, colonists were consumed by describing their environments. The birth of soil science in the homelands of the Nansemond was slavery's attempt to conserve itself against its own destruction. Though soil conservation efforts were minimal at best, the plantation economy treated land as disposably as it treated Indigenous and enslaved inhabitants. While this section does not focus on the birth of US soil science—a discipline credited to planter Edmund Ruffin's attempt to increase his plantation's yields—conservation projects predate the 1970s' efforts to preserve habitats. This is a stark reminder that the politics of conservation are located within the history of colonization's "discovery," killing, and "planting" of others, to sustain a project of white supremacy with colonial ends.[75] Attempting to improve the land, planters deployed enslaved labor to terraform wetland into terra firma (dry ground) productive for capital.

Replete with references to the soils of the Tidewater, William Byrd, governor of Richmond, Virginia, documented his 1728 journey to establish the factitious border between the first colony (Virginia) and the ninth (North Carolina). Surveying what at the time was more swamp than potential plantation, he took fastidious note of where the soil was fertile and where the soil was "poor." Though his goal was to erect a clear line of jurisdiction, he famously described the landscape as "a miserable morass where nothing can inhabit." The wetlands of the Tidewater posed a challenge to British colonists, and the suck of swampland, too, posed a challenge to the viability of their definitions of "quality of life": "Since the surveyors had entered the Dismal, they had laid eyes on no living creature: neither bird nor beast, insect nor reptile came in view. Doubtless, the eternal shade that broods over this mighty bog, and hinders the sun-beams from blessing the ground, makes it an uncomfortable habitation for any thing that has life."[76]

Contradictory to the fecundity he manifestly witnessed in the swamp— "The moisture of the soil preserves a continual verdure, and makes every plant an evergreen"—and the maroon communities nestled out of view, Byrd described the wetlands as a place where "the foul damps ascend without ceasing, corrupt[ing] the air and render[ing] it unfit for respiration."[77] Unbearable for settlement, Byrd saw the swamp as unproductive and empty. And he named the "morass where nothing can inhabit" the Great Dismal Swamp.

In William Byrd's "History of the Dividing Line," wetland and wasteland slide back and forth into each other. Less concerned with the flora and fauna, distinctions perhaps only worthy once white life is imaginable, this treatise across the swamp documents the racist eco-logics of white slave- and landowning men. The wetness of the land characterized a waste of it. While a dry piece of ground was always a description couched in deep relief and splendor, the swamp was imagined to be uninhabitable: "they had laid eyes on no living creature" and "the eternal shade . . . makes it an uncomfortable habitation for any thing that has life." These descriptions of lifelessness, or at least not a life worth living, slide noticeably into descriptions of Indigeneity: "It is a common case in this part of the country, that people live worse upon good land; and the more they are befriended by the soil and the climate, the less they will do for themselves." As opposed to reproducing more textual evidence of Byrd's racist colonial descriptions, my point here is to emphasize how the textual elision between Indigeneity and wasteland follows the settler grammar of land and becoming. Wetness slides readily into descriptions of "idleness . . . to suffer all the inconveniences of dirt," which are at times extended to the un-Christian North Carolina men of "ill fame and reputation."[78] The wetlands and the waste of them are, in Byrd's estimation, not only *dismal*; the swamp itself was subject forming, producing "*dismalites*" as a kind of uncivilized people living in wet squalor.

Though Byrd took care to mark every plantation by name and owner, noticeably absent from the racist-ecological survey are the enslaved. As Byrd notes, there was always relief provided by "the happy effects of industry [on the plantation] . . . , in which [the planter's family] looked tidy and clean, and carried in their countenances the cheerful marks of plenty."[79] Despite how much time the surveyors spent reveling in the "dry" respites provided by the plantation, the rare mention of "negro" is ecologically telling.[80] In Byrd's journey, enslaved people are unrepresentable in the visual economy of land surveys: As neither "surveyor"—a capacity of ownership that renders space legible—nor synonymous with nature itself ("the wild abyss of Native space"),[81] Black is elsewhere, haunting the edges of that which can be surveyed.

In *The Black Shoals*, Tiffany King describes how Black fugitivity and Indigenous resistance created a crisis of representation for the "White" psyche.[82] The furtive movements of those who knew the land and those who were used as a tool to transform it resisted settler incorporation in myriad ways, including through Black and Indigenous collaboration, or in the maroon settlements of the Great Dismal Swamp. The furtive movements of, in the colonist's imagination, their tools (chattel) and nature (the "savage") produced the need to rewrite, remap, and otherwise ontologize conquest.[83] Byrd's surveying project anxiously encountered native practices that he did not understand. For example, while staying with "the Indians" in the Nottaway town, Byrd writes, "Our chaplain observed with concern, that the ruffles of some of our fellow travellers were a little discolored with pochoon, where-with the good man had been told those ladies used to improve their invisible charms." Conscripting Indigenous women into continuous sexual availability, he sarcastically commented, "The Indians by no means want understanding." As the "barbarity" of unchristian practices centrally frame the Native as refusing "opportunities of improvement," including the relentless citing of Indian women's labor as that of unclean drudgery, other genres of being left settlers anxious and uneasy and in need of a straight line.[84]

Perhaps in an ironic recognition of the violence of enslavement, or as a method for making slavery and settler violence part of nature itself, Byrd explains that the "true reason" for the unchristian gendered division of Indian women's labor is "that the weakest must always go the wall, and superiority has from the beginning ungenerously imposed slavery on those who are not able to resist it."[85] Quipping about the backwardness of Indigeneity, Byrd turns slavery into an immutable fact of nature through the racist metrics that already equate the Native with what John Locke described as the "state of nature."[86] We might imagine that for Byrd, the straight lines of property and government put "men out of a state of nature into that of a commonwealth, by setting up a judge on earth with authority to determine all the controversies" and colonial governance, a solution to the inconveniences of "the state of nature or pure anarchy."[87] Byrd's description of Indigenous gender relations (Indigenous men's "natural" enslavement of Indigenous women) registers the disfigurement settler colonialism requires in order to figure white male colonists as possessing the "right to first possession" and dominion over all forms of life. Whiteness simultaneously renders itself outside and on top of nature, its devastating ecological effects thereby no effect at all. Descriptions of nature's "facts" reveal how the "natural" right of white men recursively justified conquest and murder.

Yet I also want to draw our attention to the narrative absence of the truly enslaved, indexing the impossibility of reckoning with the actual labor deployed to turn "nature" into white men's dominion. Slaves, in the eyes of the plantocracy, were tools used to produce whiteness as a form of property.[88] These "tools," though human insofar as they were punishable by lash and law, were understood to be surplus to the land. Neither fully human, manifest by the cheerful marks of plenty and industry, nor "natural" to the wetlands, the slave was spatially and ecologically unmentionable. Through its profound absence, Blackness marks a boundary between nature and white propertied life; it also marks the lines on maps through the enslaved labor that carved native territory into parcels. Racializing the slave as a way to mark who is naturally enslavable was also a method for making property a regime of *knowing* that settles matter. Whiteness's capacity to categorize, name, and survey secures terra firma as the necessary conditions of possession (dryness, cleanliness, industry). The absent Black comes to constitute "the lines" upon which Indigenous homelands were to be parceled. Within the triad where "the slave's person is excess . . . and the slave is always excess to land," Blackness is not only made ungeographic but made unecologic.[89]

In rendering Blackness unecologic, Black and Indigenous relations with the land were disfigured, obscuring how resistance and "solidarities expressed [different] land relations."[90] Indigenous resistance, including collaboration with runaway slaves, refused settler incorporation. The enslaved who stole themselves away by fleeing the plantation, whose accounts, such as Frederick Douglass's, evince "the material relations of escaping slavery," also tell a story about land.[91] In Douglass's novella *The Heroic Slave*, "the wilderness that sheltered me" expresses a relationship with land outside the relations of ownership. The enslaved also resisted the plantation's (dis)regard for land, through what J. T. Roane has called the Black commons.[92] On the plots allotted to the enslaved under regimes of "violen[t] coercion through nutrition," Roane argues, "Slaves used the plot as stolen time to engage in their own independent visions of self, family and community" simultaneously, "forwarding visions of use value despite enclosure and commodification."[93] And there were the maroons, or as Sylviane A. Diouf calls them, the "exiles" of the Great Dismal Swamp, sometimes feared for fomenting insurrection (as was the suspicion with Nat Turner's slave rebellion in 1831) and at other times described as a people "addicted to freedom," who lived with land as a means of creating life, as opposed to the plantation's violences.[94]

Against William Byrd's description of a lifeless expanse, "'the wilds' were not wild, nor were they empty—they were Indigenous homelands."[95] Nik Heynen and Megan Ybarra emphasize the importance of Indigenous studies to the project of abolition, reminding us that different "nature-society relations" have always existed.[96] "The key insight here is to understand that many communities were made through the solidarities of Indigenous peoples accepting self-emancipated slaves into their homelands, becoming peoples known today as Seminole, Garífuna, and others."[97] The independent living of those who refused to make living a value-constitutive activity for the colonies are crucial histories of other land relations that challenged, even if they could not stop, the violent alterations of human and more-than-human life that spanned the Tidewater.[98] While land surveys and the deployment of enslaved labor for ecological destruction were premised on the eco-logic that the slave was "excess" to the land and Indigeneity a mark of people "trapped in a state of nature," cultural histories of resistance, such as those described in Hosbey and Roane's "A Totally Different Form of Living," trace alternative relationships to land that challenge eco-logics forged by the settler-planters' subjugation.

The declaration that the swamp was empty provided the moral justification for Indigenous removal, birthing new projects to turn the "waste" of wetland into the firm grounds of the plantation. But wetland was not ideal for the tobacco and cotton plantations that colonists so deeply desired. So not long after Byrd declared the swamp *dismal*, George Washington, originally a surveyor by trade, created the Dismal Swamp Company, which sought to drain the swamp and fill it with plantations.[99] From 1764 to the early 1770s, enslaved labor was used to dig more than 150 drainage sites to drain the swamp. Though the drainage was moderately successful, Washington would learn that the loose wet peat soils of the swamp could not be cultivated for tobacco or cotton.[100] Thus, the land company quickly became a timber company, employing "able bodied slaves" as shingle getters.[101] Continuous logging created sediment problems, altering delicate ecosystems and clogging waterways. Though George Washington's company failed to turn the swamp into plantation lands or even turn a profit, settler colonialism and chattel slavery had an impact on the swamp's political ecology, which would manifest itself in the future in the floods and fires that plague the swamp today.[102]

The swamp, estimated preconquest to have ranged from about 1 million to 1.4 million acres, is now one-fifth of its former size. The ecological effects of slavery are visible through the depletion of biodiversity, as noted in the absence of the great panther.[103] These effects were also visible in 2011, when lighting struck the now-too-dry peat soils in the swamp (no longer able to absorb floodwater), starting a wildfire that burned for 111 days.[104] And slavery's impacts are visible in the centuries of logging that brought development closer and closer to the outskirts of the shrinking swamp, places now plagued by the effects of deforestation and erosion of the sandy loam soils now home to residential development. According to a US Geological Survey report done in partnership with the US Fish and Wildlife Service, "Timber harvesting and the construction of ditches to drain the swamp [which] facilitate [timber] harvesting are collectively implicated in changes that altered the wetland forests."[105] Routinized racial violence is not nearly as immaterial as white supremacy often makes it seem.

As previously mentioned, during the conservation movement of the 1970s, the Great Dismal Swamp was designated as a wildlife refuge under the purview of the US Fish and Wildlife Service (1974). Environmental projects attempting to rewet the swamp "intended to re-establish the original wetland-forest types, reduce the risk of fire, reduce subsidence and decomposition of the peat, and enhance peat accretion."[106] Such projects to "restore" the swamp as a refuge for wildlife have yet to imagine themselves alongside, in the name of, or within the cultural and political purview of those who made this place a refuge from violence. The only acknowledgment of this history is a platform that sits about half a mile into the "trail" that Jane told me to walk at the top of this chapter. A circular plaque reads, "The Underground Railroad, A Network of Freedom," and underneath it three panels attempt to summarize "resistance and refuge" from 1680 to 2014. From fugitives to commercial loggers, the panels present a story of freedom in which enslaved fugitives were eventually liberated into the waged labor of commercial logging. Perhaps ironically, one panel claims, "The US Fish and Wildlife Service has used logging as a tool to improve habitat for wildlife."[107] Though the panels do acknowledge that after the Civil War, vagrancy laws and the Black codes *required* the formerly enslaved to find waged work, the story stops short of asserting that the subjection of enslavement was an environmental problem.

In addition, the panels make no mention of the Nansemond Indian Nation and the Algonquian speaking tribes that made up the Tsenacomoco alliance of Chief Powhatan, who not only dwelled here but also refused to trade with the English settlers.[108] From the first Anglo-Powhatan War (1609–14) to the third Anglo-Powhatan War (1644–32), the Nansemond were forcibly moved time and time again as Europeans settled along the Nansemond River that once fed into the swamp. In 1792, the Nansemond were forced to sell the land that the English had "given" them, making it the tribe's last known reservation. According to the dubious records of colonists, "Only three non-Christianized Nansemond survived [with] the last [known member reported to have] died in 1806."[109] But not unlike the present, where white claims to Indigenous ancestry are a settler nativist "move to innocence," the Racial Integrity Act of 1924 accommodated elite Virginians who claimed Pocahontas and John Rolfe as ancestors by defining those who had "one-sixteenth or less of the blood of the American Indian and ha[d] no other non-Caucasic blood" as "white persons," erasing the Nansemond as a category of people under Virginian law for whom "records of their existence" could be kept.[110] The Nansemond tribe was not recognized federally until January 29, 2018.

To be blunt, the designation of "refuge" remains an all-too-ironic reminder that the alteration of the swamp's ecologies was predicated on the violent eco-logics of setter-colonial plantations. While Indigenous and enslaved people's intimate knowledges turned the Great Dismal Swamp into a refuge from violence, the epistemological and ontological scales of genocidal parceling transformed the material conditions of possibility that produce the present ground. Thinking about the violent regimes that define who and what is made "excess" to land opens up the possibility to ask new kinds of questions about the ecological past buried in the human and more-than-human present. These historical logics offer source points for attending to the *longue durée* of un/freedom and the ecologies of Black becoming with altered lands.

As the surplus to capital, waste is often presumed to be the "end" of the commodity's life.[111] But calling something waste (as in the declaration of Indigenous homelands as wastelands) inaugurates world-ending beginnings. The commodification of waste is an enclosure taking part in the surplusification of life, and surplus life is being reserved for future *material* uses, perhaps to site the ever-expanding need for landfills that capitalist-colonial relations have wrought. The end of the world is not an event to come but an ecological condition of past futures filling in the ground of the present. Black life is becoming fill as houses are sucked into the wet ground, as "excess" Black life becomes fill when a landfill extension requires it. The unwaged labor forced to drain wetland was built on an imagination that land could be filled with the plantation and, in so doing, produced the ecological condition of Black wageless life becoming fill in the toxic capture of the Black present. Becoming fill marks the extension of the unecologic to the transformation of waste into a commodity, and it reveals the way that Black un/freedom endures through matter. But it is also, perhaps ironically so, a reminder that Blackness is connected to land. Histories of resistance reveal alternative land relations papered over by the racist logics of land use that bring the past in close proximity to the present. Still, I want to hold onto this eco-logic for its lessons: Rendering Blackness excess to and for land (dis)figures Black people as a threat to property. And perhaps, in the spirit of abolition ecologies, we ought to be.

SURPLUS

Waste keeps on accumulating and it's always *on the move*. It flows from New York City to Virginia's landfills down I-95. It flows through the streets with the help of wind and feet. It flows through neighborhoods on management routes carefully coordinated by municipal solid waste trucks and private haulers. It flows across state lines to transfer stations, where it is sorted onto bigger trucks called yard dogs that cart waste to recycling plants or treatment sites or material recovery facilities (MRFS) or landfills. And then waste matter transforms. At the MRF, some objects are melted and others become ash. At the landfill, air and pressure turn putrefying waste into a thick liquid substance called leachate that needs to be captured and treated before being injected back into water. Methane escapes, adding to the CO_2 emissions that slowly warm the earth, and the warming atmosphere, which can hold more and more water, makes landfills produce leachate at faster rates and in greater volumes. And then, of course, there are the commodities that with every "greener" intervention produce new contexts for new toxicants, generating the need both real and imagined for land to "host" more pollution. Landfills expand, MRF and waste-to-energy facilities become overtaxed, and this networked technocracy expropriates more and more land. What is waste management but the management of surplus matter and surplus geography?

MATTER

The contract that forever linked Virginia to New York City's discards was merely the beginning of what can only be described as a landfill boom. In the span of three years, seven mega landfills were built in Virginia, ushering in unprecedented waste. It might seem as if the 1990s landfill boom was due to an uptick in surplus matter. The massive flows of waste that began heading south seemed to be on a scale proportionate to population growth, a surplus production of *things*, an accumulation of what many environmentally informed political pundits describe as the "rampant consumerism" that plagues us all. "Trash, a pollutant that arises daily and visibly in the household, is always at hand as an object upon which to work up environmental concern and action in the realm of everyday life." Some environmentalists, not those who live next to landfills, of course, tend to link global waste problems to consumerism, producing the attendant belief that consumption can be ethically reformed, in particular through recycling. Samantha MacBride, however, forcefully argues that these arguments eclipse the massive amounts of waste produced at the point of *production* and frustratingly obscure the way recycling itself produces waste.[1]

Over the last ten years, news outlets have brought public attention to the failure of recycling facilities in the United States. According to a 2022 Greenpeace report on the state of recycling, the vast majority of recyclables in the United States are landfilled.[2] MacBride's work argues that the United States does not have the infrastructural capacity to recycle as much recyclable material as it generates, and so it "outsources" the recycling of particular materials (such as cloth) to the Global South. But the EPA's reporting on waste cleverly disguises this outsourcing as "recycling." For example, in 2018, the EPA reported that the United States recycled 3.09 million tons of material; however, 1.04 million tons of US plastic waste was actually *exported*. Moreover, most plastic objects that are thrown into recycling bins—such as soda bottles and plastic food-service items—don't meet the threshold of "recyclable material" as defined by the Ellen MacArthur Foundation's initiative for "determining whether a plastic product is recyclable."[3]

Thresholds and definitions of an object's capacity are part of the (in)ability to (ac)count (for) surplus matter. Recycling rates and waste tonnages are numbers that are as elusive as they are revealing. These numbers tell stories that obscure what counts as recycling as well as what, who, where, and how waste is actually managed. For example, the difference between how garbage bins and recycling logos dictate responsible "environmental" behavior is radically separate from what an MRF will accept. Moreover, just because an MRF accepts an object does not mean that it will be recycled, especially if there is no market for it.[4]

Going beyond the story that Greenpeace tells with numbers, MacBride notes that on top of the fictional solution that recycling has been used to promote, recycling has also produced a moral and political economy around *particular* materials, politicizing some objects while ignoring others and their complex locations. Textiles "have serious environmental impacts at all stages of the lifecycle," but they also constitute an important "secondary market" outside the United States.[5] Similarly, scholars writing on electronic waste (e-waste) note the increasingly complex mobility of e-waste and its impacts on the Global South.[6] Geographer Rajyashree N. Reddy, for instance, writes about the way "multidirectional flows . . . converge to make Global South cities vital conduits and nodes" of extraction, turning places like Bangalore, India, into urban mines.[7] The mobility of wastes and its confluence with race is "situated precisely where the geographies of social environmental inequality that design the world dictate that it ought to be."[8]

The politicization of *particular objects themselves* conveniently sidesteps a larger problem: "Waste [is] not a 'thing,'" Marco Armiero has said; rather, it is a set of "wasting relationships."[9] To that end, Armiero and Massimo De Angelis stress that we need to shift our attention toward "the contaminating nature of [racial] capitalism and its perdurance within the sociobiological fabric, its accumulation of externalities inside both the human and the earth's body."[10] This is part of what Simone Müller writes about when she argues that "toxic exposure has become—while unequal—increasingly common"—or, simply put, a "toxic commons."[11] While a toxic commons might be a generative place from which to reimagine our relationships to toxicity, specters "of excess" haunt the social relations dictated by racial capitalism and, in particular, how its surpluses—including the surplus of poor disposable life—threaten its production of *order*.[12]

"Poverty," Gidwani and Maringanti write, "becomes civil society's itch, a source of indeterminacy to its propertied citizens."[13] While the proletariat in Marxist terms is the titular revolutionary class, the wageless poor, specifically, the lumpenproletariat, is a source of anxiety even for Marx and Engels: "The passively rotting mass thrown off by the lowest layers of the old society, may, here and there, be swept into the movement by a proletarian revolution; its conditions of life, however, prepare it far more for the part of a bribed tool of reactionary intrigue."[14] For Marx and Engels, the lumpenproletariat was, in some sense, too outside capitalism's central antagonism—the bourgeois versus the proletariat—and thus fundamentally disorganized for the potential for revolutionary consciousness. Yet the lumpenproletariat is a class that, as Gidwani and Maringanti note, prefigures the importance of "surplus populations" that seem to anxiously gnaw at both the working-class and the owning-class sense of order. Thus, Gidwani and Maringanti insist on the sustained attention of the labor of surplus populations as *infrastructural* in order to irritate "the logic of capitalist value making."[15] This sustained attention prioritizes the recognition of the "*noncapitalist* forms of value making that give substance to social life."[16]

The conditions of waste management are toxic, and the regulatory apparatus upon which an imagination of ethical consumption relies entrusts capitalism with the meaning of matter. Waste is managed on a scale that, in daily life, even for those who live near a landfill, is incommensurate with the enfleshment of its burden. The obsession with consumption is part of a neoliberal ideology that believes that individuals can and should take responsibility for the way racial capitalism *wastes*, while simultaneously entrusting a deregulated market and defanged regulatory apparatus thoroughly overrun by capitalism's demands regarding the definition and *value* of matter. But if surpluses of matter and people anxiously frustrate racial capitalism's order, why not center them as sites of resistance—particularly resistance to extractive regimes of value?

GEOGRAPHY

The massive flows of waste inaugurated by the rapid construction of seven private mega landfills in Virginia in the 1990s created a crisis of capitalist competition. Directing the flow of waste directs the flow of profits. Flow controls are the "legal provisions that allow state and local governments to designate the places where municipal solid waste (MSW) is taken for processing, treatment, or disposal."[17] Ostensibly a provision intended to protect the environment as a public good, flow controls don't so much protect the environment as much as they determine *where* competition happens. State control over the flow of waste is at odds with the "spirit of the market," making waste regulation a site around which the waste industry and local management produce their own geographies.

In 1976, Congress passed the Resource Conservation and Recovery Act (RCRA), which sought to regulate the disposal of solid and hazardous waste. However, it was not until the 1984 Hazardous and Solid Waste Amendment, which expanded the EPA's oversight of state waste infrastructure, and the 1992 Federal Facilities Compliance Act, which required that waste plans be overseen by regulatory bodies, that trash management was asked (at least legally, if not in actuality) to clean up its act. The new regulations required states to update municipal dumps to sanitary landfills, requiring liners, leachate monitoring systems, and treatment sites, and creating protocols to cap landfills with clay or ash. These new protocols and technologies were costly, sometimes totaling up to $40,000 per acre. In the absence of redistributing resources to redress sedimented toxicities, environmental regulation did little more than open the floodgates for the private waste industry. EPA compliance caused a cascade of dump closures, creating a multitude of Superfund sites (also to be overseen by the EPA) and a reliance on private haulers, who, in the wake of dump closures, realized the decentralization of waste management had created a new profit center: managing waste across distance.

Sanitary landfills are capital-intensive projects. As the previous figures suggest, updating dumps to sanitary landfills was often more expensive than closing dumps altogether. When over 240 facilities closed in Virginia in the 1990s, municipalities began to turn to waste industry giants such as Browning Ferris Industries, Republic Industries, and Waste Management Inc., which were setting up shop in the cheap sprawling land of the post-plantation South. After the 1980s saw a rapid decline in the value of farmland, waste companies began securing cheap (often overfarmed) land for the construction of mega landfills. For the same reason that cheap land became attractive to the waste industry, it also became attractive for the construction of supermax prisons.[18] This new use for fallow land changed the way and where of waste flows. The scale of these landfills not only allowed private companies to contract with municipalities at a lower cost but allowed companies to contract across larger distances than the small private haulers who trucked waste within the boundaries of Virginia. The ability to truck more and across longer distances made it possible for private waste contracts to become longer (both in time and space) and more complex.[19] While prior to the 1990s private haulers were typically responsible for commercial waste and municipalities for residential waste, the birth of the private waste industry afforded a cheaper option to cities than making their local dumps EPA compliant.

Though the expansion of the EPA's purview sought to address the deleterious effects of unlined dumps, the regulation of hazards did not come with any funding to do so. In fact, the EPA's attempt to promote "human health and the environment" in the absence of resource redistribution *promoted* the marketization of waste, giving waste corporations increasingly more authority over the management of "the environment." The confluence of stringent waste regulations and the privatization of waste management created a "reconstitution of state-economy relations," including endowing the waste industry with geographic control over "market-based regulatory arrangements."[20] These geographic and economic arrangements promote the health of capital's accumulation. Market arrangements don't benefit the already economically disenfranchised, nor do they protect the ecologies of land and living, as examples from the water crises in Flint, Michigan; Newark, New Jersey; and Cancer Alley in Louisiana show. Moreover, market-based regulatory arrangements tend to authorize the geographic vulnerability of those *already* vulnerable to other forms of risk, turning relationships among people, land, and industry into an exchange of health *for* capital.

Although environmental scientist and former Virginia Department of Environmental Quality worker Vivian E. Thomson notes that some of these poor communities have invited these industries into their homes, she argues, "this is the exception rather than the rule." When "home" or "health" is turned into something to be exchanged within the waste industry's geographic control, it is more accurate to speak of racism's coercive logic of slow death *as* the health of capital.[21] Yet it is equally true that waste has the potential to threaten capital's accumulation. "As history's eruptions have shown repeatedly, the things, places, and people remaindered by the establishment can never be fully sublated."[22] Capitalism's constant production of surpluses might sustain it, insofar as capitalism is an economy predicated on the creative use of surplus and labor to produce value, but its inability to fully incorporate the waste that its wasting relationships produce is an important distinction to hold. And we can hold on to this knowledge even as we notice how "capitalism can selectively subsume informal waste economies, as, for example, ongoing efforts by corporate capital to capture wealth in waste by entering the terrain of 'waste management.'"[23]

3 Revisions from Elsewhere

Where there is dirt, there is a challenge to the system.
— LIBOIRON AND LEPAWSKY, *DISCARD STUDIES*

Toxic capture makes the materiality of Black life punishing. When signs of "decay" are the same signs that inspire excessive police force, the criminalization of poverty, and state-sanctioned murder, anti-Black violence not only buttresses Rashad Shabazz's insistence that "public Blackness [is] akin to a crime" but reinforces *punishment* as a central ecological relation.[1] For example, "Prisons, detention facilities, refugee camps, and violent borderlands function to *concentrate* surplus life."[2] These geographies of punishment inform how Black people *ought to* relate to land in this settler imagination—Indigenous to nowhere and, as such, a menace everywhere. This US spatial logic is also an eco-logic. Black people are relentlessly cast outside nature, as a contaminant to land and a threat to property. Blackness's fungibility is not only the product of an arithmetic forged by chattel slavery but an ongoing relationship to the injury of racialization, which, according to Willie Jamaal Wright, "is connected to land and the environment."[3] As I have shown in chapter 2, this ecologic relation apprehends Black people as *un*ecologic in order for settler-colonial racial capitalist land relations to sustain themselves, and it does so through the mundane toxification, surplusification, and discarding of Southern Black life.

In this chapter, I start again in the South, but bring the reader with me back north, to New York City, where toxic capture produces wageless and houseless Black people as unecologic too. Albeit under racial capitalism's differential Northern geographies, where surplus land is less likely to be plantation-toxified soil, and are more likely to be the broken windows of white supremacist capitalism's accumulated (post)industrial (and poisonous) material violence. In New York City, where Blackened neighborhoods are relentlessly subjected to accumulation by dispossession, Black people, yet again, are held captive as a form of environmental control. Up North, historically Black neighborhoods, where waste is presumed to be a problem of the hood and gentrification (and policing) a sanitizing solution, I work to show how criminalized and houseless Black scavengers in Brooklyn advance ecologic modes of telling and living that refuse to be cast out of the city's ecologies. Against the forms of dominion that discard the poor, discarded Black living challenges the nature of property's value(s) and property as natural.

While the "unecologic" describes racial capitalism's *use* of Black bodies, it does not describe the way Black people relate to capitalism.[4] Instead, my use of the word is to insist that it is an outcome of the "settler-native-slave" triad, in which "the slave's *person* . . . is the excess" and "the slave's very presence on the land is already an excess that must be dis-located."[5] Adding to the chorus of scholars who think through this Black inheritance, a subject position structured by objectification and a personhood (or humanity) structured by punishment, the unecologic names one of those ways that racial-capitalist settler colonialism persists.[6] Relentless in its determinations of *genres of being* that menace the white settler-owner's domination—marronage, Indigenous sovereignty, houseless life making, and more—other ways to be are "reconfigured/disfigured as the [killable] threat."[7]

For the purposes of my ecological argument, we might understand toxic capture as a form of disfiguring (meaning to render maimable and killable) that articulates an important alignment with what Black disability scholar Sami Schalk calls the disabling "material impact of racism" and M. Murphy calls the accumulated harms and complex collectivities of alterlife.[8] The punishing materials that circumscribe Black life recursively justify the reign of private property, and so too do they justify a narrow vision for theorizing material relations disfigured by theory. Wealth accumulating "here" and toxicity, waste, and risk accumulating elsewhere materialize dominion as the white sovereign common sense.[9] Thus, it cannot be overstated that the unecologic (or what I sometimes refer to as "this eco-logic of property") is a theoretical proposition and one that must not be confused with Black environmental projects, Black ecologies, or Black life itself. In fact, the unecologic is part of the grammar of (whiteness as) property that makes Black environmentalisms illegible but that Black livingness makes abundantly clear.

The places produced by or shot through with the violence of toxic capture might be elsewhere, but they are also *everywhere*, and they are, most importantly, places from which poor Black environmental practices spring. These places are multiple, like the Great Dismal Swamp where the colonial resistance to Black freedom bears out the ecological consequences of the plantation, or the "Black" zip codes where lead levels are high and the struggle to breathe is turned into a problem of Black behavior, or the so-called blighted neighborhoods produced and then disappeared by the redlines of property values. These places—places of unruly entanglements of devalued matter and devalued life—are places filled with Black people, whose very *living* is an ecological achievement.

Instead of trying to satisfy the desire for a programmatic and scalable vision that leads to a sustainable future, this chapter urges us to read *against* the anti-Black eco-logic of property and to revise the standpoint from which the environment is thought about. Instead of thinking *about* the environments of toxic capture—those places we might call "blighted," "empty," "trash-filled," and sometimes "wild," "unruly," or "dangerous"— what if we thought *from* them? What if, instead of thinking *about* waste, we thought *from* the locations, practices, and modes of existence that adhere in being discarded, rendered "excess," or "unecologic" as environmental projects in and of themselves? Thinking *from* requires an epistemological shift that centers a "positionality . . . formed in concert with spaces of unvalue," those "un-" (or, to my preference) "in-" credible authors whose survival is criminalized and not thought to be ecological at all.[10] What I hope to suggest is what I was taught: Spaces of unvalue are formative to the modes of *being* that might be "a release—a mode of existence beyond the torments, logics, and habits of" settler-colonial racial capitalism.[11]

ARCING CLOSER TO THE DIRT

This chapter centrally flips the lesson that Mary Douglas taught us long ago, that dirt is matter out of place.[12] Instead, I urge us to arc closer to the dirt/trash and to claim it as a Black feminist (and Black materialist) politics. Many scholars of color have rightfully focused on the work that *dirty* does as a denigrative description of both matter and body.[13] These scholars have been central to producing an intersectional program for calling out the racism of heterosexism and the "color" of purity. In addition, Black feminist scholars have pointed out the way that *dirty* signifies on *savagery* as Black persons too close to the jungle, so to speak.[14] While these interventions shape my own analysis—and certainly describe part of my own experiences as a Black woman *stained* by the violent histories through which I was made possible[15]—I want to shift our attention to degraded matter and the names we give to it ("dirty," "trashy") to think differently about race, gender, and materiality.

Dirty and *trashy* are descriptions of a state of matter, but they are also descriptions of people. They are racially and sexually charged ascriptions that in their naming both proffer a state of contamination (something that has already come to pass) and mark people as contaminating (something that names their present and future danger). Ideologically, as a description of an inferior (present, past, and future) state of being, the ideology of dirtiness works to control the purity of the future by condemning present sexual relations. In the United States, this has long been the providence of miscegenation laws that sought to regulate the contagion of mixing with Black blood. Tribal blood quanta, though complexly important to Indigenous resistance, is also a way of regulating sexuality though notions of purity and contamination.[16] In addition, ideas of purity and contagion are central to caste systems, in which low-caste people's untouchability prescribe a place in the political economy, eliding Dalit women's political claims—namely, to land.[17]

So why arc closer to the dirt if it is a place where race, class, gender, and sexual domination is exerted *materially*? Moreover, why see this as a materialist feminist project, and a *Black* material feminist project at that? The late bell hooks reminds us, "Feminism is [not only] a struggle to end sexist oppression . . . it is necessarily a struggle to eradicate the ideology of domination that permeates Western culture on various levels."[18] Racist, classist patriarchal domination is spatial; dominion enrolls our material world in a project to constrain, dominate, and enclose the criminalized, pathologized, feminized "other." How people make life with those material constraints must be critical to our reconceptualization of environmental justice. This means including those people who are relentlessly written out of history and our contemporary environmental politics: drug users, sex workers, the poor, the houseless for whom waste matter is what is used to survive—and even thrive.

Black ecologies scholars remind us that fugitive slaves used the unruly "wastelands" of the swamp to craft spaces of freedom.[19] So, too, do the criminalized poor, caught within the condition of toxic capture, use the waste of the city as a material reality through which to chart lines of flight. And these lines of flight challenge the spatial dominion upheld by property relations. Discarded and criminalized people use objects to (re)compose and curate space, to wrench open a place to *be*, even if, or especially if, those object relations—like the "theft" of recyclables or the crafting of a phone while incarcerated—are violently sanctioned by the state. At times, this place reveals other ways of thinking gender and at others reveals how racial objectification, that is to say how property relations and racism turn Black people into objects, makes for an alternative relation to our material world. This place reveals a different horizon of ecological ethics that affiliates Blackness with pestilence and femmeness with trashiness, rejecting a genre of being human in which distance from dirt marks you as belonging at the top of the great "Chain of Being."[20]

In arcing closer to the dirt, I want to bring Black feminist thought to what feminist science studies scholar Amade M'charek calls "the art of paying attention"—an art that requires attention to how power arranges matter and which matters matter to powerful arrangements. But it also requires thinking differently about how object use produces "the capacity to know."[21] Gail Lewis writes informatively about the way Black feminism produces objects through practices of curation.[22] Part of the importance of Black feminist thought, she argues, is its ability to continuously compose objects that, in relation to me (subject) and Black feminism (object), facilitate a process of Black feminist *becoming*. Though Lewis is talking about Black feminism as an object composed of articles, special issues, anthologies, and edited volumes curated by activists and scholars to produce a field of play, I'm interested in thinking about how we might honor a kind of dirty curation (improvisational practices of assembly) to think about Black feminism. In other words, how does Black feminist thought help us think differently about materiality and object use as part of the stuff that facilitates Black being?

This is necessarily a *materialist* project, a mandate to stick close to the materiality and materials with which Black life is made. But it is also a Black feminist materialist project that does not seek the universality of humans to locate alternative ecological politics.[23] It does not presume to think from the nonhuman but rather understands *humanization* as a violence of Western epistemologies.[24] This materialist project sticks close to the way racism uses matter (which includes flesh as well as material qualities and capacities of objects) as much as it is concerned with different ways of knowing matter that facilitate alternative human, land, and more-than-human relations. Also, it includes naming the way "gender is a racial arrangement" of materials (which is both historical and geographic) and theorizing the way racial, classed, hetero(cis)sexist objectification makes objects available for commodification *and* cultural production.[25] While in many ways the racial arrangement of materials and the materiality of racism have been central to the way this book moves through place, time, and practice, racial and hetero(cis)sexist objectification needs further elaboration.

(FEMME) DEVIANCE AS RESISTANCE, OR TRASHY POLITICS

First and foremost, in arcing toward the dirt(y)/trash(y), I draw on L. Horton-Stallings's theorization of the "fecundity of 'dirt'" as a "terrain of sexuality and gender studies."[26] Feminists of color have detailed how patriarchy imposes gendered degradation through racial objectification.[27] But more than architectures of domination, Black feminists have sought to strive for reparatory readings that insist on the plentitude of Black women's perspectives and analysis of resistance and liberation through time and space.[28] One such place is by locating Black feminist thought in Black cultural production.[29] While some have challenged hip-hop, rap, and the aestheticization of Black culture as part of a liberatory project, others locate epistemological critiques in the way Black people make place and time for themselves with/in the Black material cultures demonized as "deviant."[30] For Horton-Stallings, the genre of Dirty South hip-hop proliferates a dirty imagination that mounts an "aestheti[c] and artistic critique of moral authority . . . whose purpose is dismantling and reinventing . . . the sexual economy of slavery . . . sustained by settler colonialism."[31] While Horton-Stallings is importantly locating this practice and politics in "southern public spheres," I see "'the dirty,' . . . in its finest and filthiest iterations," as a way to notice the "place and practice of intersectional politics" in the many Black elsewheres made and (re)made with matter and alternative environmental imaginations.[32] To that end, this chapter jumps across time and space, making connections between the different ways that Black wageless life is lived. Dirt and trash are inexorably part of discarded living, and in its finest and filthiest iterations, discarded living *is* an environmental/ecological critique. Following Stallings's dirty imagination, I take seriously Cathy Cohen's call for a queer Black feminist attention, "a new analytic framework for the study of Black politics, that of deviance."[33]

Deviance is dirty, if not downright trashy, in more ways than one. And the long history of degrading Black flesh is critical to theorize if we are to follow Cohen's insistence that there are "liberatory aspects of deviance."[34] To put it differently, understanding the lines of Black femme flight between the degradation of matter and the degradation of Blackness is one of the places where we might locate what Harriet Jacobs called a "loophole of retreat."[35] My use of Black femme flight comes directly out of the work of Treva Ellison, whose analysis of the trans and queer geographies of Black gender and sexual nonconformity defines this line of flight as "the re/appearances of queer femininity that disorganize and confound the categories we often use to make sense of the world."[36] While Ellison develops this analysis by rereading the transphobic inscriptions of jailed Black femmes, I deploy Ellison's analytic to remind us that disorganizing and confounding categories need to be extended to the presumed stability (especially the way we *name* that presumed stability) of our material worlds. As Black trans and queer scholars remind us, "gender and sexual non-conformity has historically been affixed to Black people and places" and is part of the hetero(cis)sexist racial economy of Black degradation.[37] To be clear, my interest is not in the assumption that Blackness is degraded but rather in paying attention to the loopholes produced by *making use* of degradation's (mis)names to live differently. A Black femme line of flight might just reveal a horizon in which discarded objects and discarded living turn the material world into a field of play—a field of play that Horton-Stallings might call fine and filthy living.

Overburdened by the racial-sexual economy of condemnation and "the wounds it inflicts on black women's flesh," Black feminist scholar Jennifer Nash insists that an attention to Black ecstasy (particularly on that which has long troubled Black feminists, racialized pornography) forces us to also look at the ways that Black feminism approaches representation.[38] Whether they are in pornography or enrolled in the representation of diversity, Black women are disciplined by the mandate to be respectable in order to be representable. Drawing on Evelyn Brooks Higginbotham's work on antiracist activism and the immense "discursive effort of self-representation" for Black women reformers in the early 1900s, Evelynn Hammonds writes that no matter "the truth of their lives," Black women's rhetoric chose "the politics of silence" to construct a respectable sexuality.[39] The *use* of Black women's bodies—as gynecological laboratories, as testing grounds for juridical law, as zoological displays, and as factories for the production of commodities in the form of humans—makes little room for an "agent" to emerge, much less be represented. Thus, representation (and Black feminist critiques of it) often (mis)name the conditions of possibility for agency, including the agency to remake gender and find what C. Riley Snorton might call the trans capabilities within the un/gendering of Black flesh. I read Nash's addition to our Black feminist vocabulary as a demand to read differently; to read in ways that allow oppressed people to *act*, *think*, and *desire* agentively—defiantly and subversively. Moreover, it reminds us that within the interstices of racial and hetero(cis)sexist objectification, people wrench open the disfiguring relations of racial capitalism's death-dealing to make life on their own terms. Black women, especially when poor, queer, and trans, are so trapped by the materiality and theorization of condemnation that their (mis)names, Hortense Spillers has said, "are markers so loaded with mythical prepossession that there is no easy way for the agents buried beneath them to come clean."[40]

So, what if they didn't?

Following the lines of Black femme flight from degraded matter to degraded flesh just might lead us to an abundant horizon of ecological politics. Paying attention to the deviant practices of object use that critique racist heteropatriarchal class relations leads us to the trash and the improvisational forms of curation that make Black being possible in its finest and filthiest iterations. To say this differently, a close attention to the lines of Black femme flight might just bring us closer in line with the material worlds that have long come into being—a vexed state of waste and/as life, the disposability of land and its continued devastation, as well as alternative fields of playful being forged against the violence of (settler-)colonial racial capitalism. In this chapter, people find affinity with waste and pestilence, not distance from them. Within the regimes of domination that Max Liboiron and Josh Lepawsky argue must discard in order to maintain power, it's all the more important to insist that "where there is dirt, there is a challenge to the system."[41]

TRASHINESS AS FEMMENESS

Criminalization is a convenient way to police Black being and reinforce anti-Blackness as an environmental strategy. The carceral logics that constrain when, where, and how Black people move use signs of "decay" to contain Black life elsewhere. In Virginia, where two-thirds of New York City's thirteen thousand tons of solid waste is diverted, the reuse practices of houseless Black sex workers are equally constrained.[42] Treated as "a human *and* environmental hazard," sex workers can be removed by panhandling ordinances, and sanitation workers treat their presence similarly to the way they treat graffiti: as "blight."[43] Since 2009, WMI has been training their garbage-truck drivers to "act as an extra set of eyes and ears for sheriff's deputies, police, fire department and emergency services, reporting emergencies or any suspicious activity that may take place during their routes."[44] In an all-too-articulate alignment of sanitation as/and policing, Waste Watch—WMI's new community partnership with local police—makes waste removal and the removal of criminals one and the same infrastructural task. Again, the goal is the protection of white property. Property is a white supremacist force that insists on particular object relations, including the spatialization of gender to *naturalize* colonial habitation. Criminalizing nonpropertied wageless life is a way to sanction the ecological ethics that mount a totally different way of living, including trashy women living with waste.

In the Tidewater, where I found myself in conversation with self-described "trashy hookers," I learned how to chart lines of flight across criminalization's mandate that the wageless Black poor are a threat to "the environment"—in other words, a threat to a propertied and colonial way of life. Though the white gaze criminalizes unsanctioned waste relations, discarded Black women gaze back, trashing the horizon and ushering forth the elsewhere as abundant matter and Black feminist ecological thought. As I noted in the introduction, this book disorders the space-times through which we come to think *with* waste. This section focuses on restaging waste as a scene of alterity that "rescripts how Black female bodies move" to see the ecological art of living in unforeseen places.[45]

Before I was invited into the fugitive living room that the girls crafted in the parking lot, before the girls would jump into my car after long days of failed fieldwork at the landfills and say, "bitch where you beeeeen I need some Mc-y D's!," before I was one of the girls to whom the mutual aid of discarded living was extended, and before I realized that these "trashy-ass hookers" were teaching me more than I could thank them for, I was a stranger at a motel in Virginia Beach that seemed to collect geographically condemned women. Eventually, one of the girls approached me and struck up a conversation. At first, I thought it was a courtesy in exchange for a smoke, but I quickly found myself amid what felt like a barrage of questions—who are you, what are you, why are you here? Betty was suspicious of my answers—a graduate student from New York City doing research. "Hmm . . . New York, I guess that makes sense. You kinda look like a city girl. Still doesn't really answer how come you ended up here. . . . You hookin'"? The question made me laugh. Betty was brazen, or perhaps just straightforward. And I liked it. Admittedly, the fact that she looked like a fly girl—a kind of Black/Latinx working-class aesthetic that my Black immigrant mother insisted I not partake in so as not to mark myself as *that* kind of Black girl—was also a factor.

"No. I'm just trying to find my way down here. New York sends most of its trash down here, and I guess I'm just trying to find it"? As I spoke, her eyes wandered: watching my fine but tightly curled hair move, looking at my two-tone lips change shape as I spoke, cocking her head to peer at the palimpsestic arrangement of British and Dutch colonial histories of African enslavement and Indian and Chinese indentured servitude in the former colony of Trinidad and Tobago etched into my face. She scanned my features—something I am used to as a transplant from the Caribbean, where the phenotypical expressions of Blackness provoke question.

In the early nights of ethnographic fieldwork, Betty told me stories that (re)mapped the urban Tidewater—hookin' in Norfolk, visiting family in Suffolk, doing drugs in Portsmouth. But these maps, I realized, were also a way to ask more about who or what I was. She had already declared what *she* was, "swamp trash"—a Black geography, a Black history, a Black ecology, and perhaps even a Black sexuality—and New York, at least for her, was not a landscape endowed with poetics that could clearly mark what *I* was. That I was "looking for trash" only further obscured the class markers I imagined Betty to be searching for on my body.

"But you know," she said, "that's a strange thing for you to be looking for . . . out here . . . alone. You really down here *alone*? You're so pretty. You have such nice hair." It had been a few conversations like this before Betty asked for another piece of information: "No man by your side?"

I shifted a bit in my stance. This was a moment that I knew I would have to confront but had not yet rehearsed. And over time, the more adjacent I realized "the field," in anthropological parlance, was to the fields of the plantation, the more I invited what I was about to say. But I hadn't yet heard the possible forms of epistemic liberation echoing across Betty's Black geographic invocation—the swamp houses' local histories of marronage that plot, to invoke Justin Hosbey and J. T. Roane, the cultural histories of other ways of living.[46] I also hadn't yet had to (or perhaps it's more like *wanted* to) correct my presumed sexuality against the straight lines of patriarchy's racializing sight: "No. I mean, I have a partner . . . but . . ." In my hesitation to define myself as straight, a world of possible readings opened up, none of which removed patriarchy but rather revealed the crook in the not-so-straight ways Black women come to see each other in a complex geometry.[47]

Betty saw me anew. "Oh! That makes much more sense. You're *not all about that dick*. I got you. Me neither, girl. This man I'm traveling with now is a bum, but you know, he got that car. We sleep in it sometimes. Yeah. I see it now."

I was relieved but also confused. "See what?" I asked.

"That you'd like a *stud!*" she said, then continued:

> You got this out here. Yeah, you'll be aight. But no one is gunna understand you, you know? It's not like people be beatin' up on you or nothin' cuz you're pretty. They won't touch you; I don't think. As long as you stay away from those country n****s but you know, you just don't look like you *belong . . . here*. It's like you don't quite look, you know, like you're all about that dick-needing: No man and having his babies, caught up on welfare and all that shit. You look too classy for that. But not like stuck up. Just different. And you're out here looking for trash? You like to get *dirty*. Yeah, I get it, I see you. I see it now, you [trash] too!

As indicated by the bracketed insertion, Betty didn't say "you trash too"; she just said, "you too." This revision, incomplete and, some might say, perhaps unnecessary, marks a moment of expansion, where the field opened onto another vision: a calling in*to* something else.

That Betty could be saying "you're trash too" or "you're queer too" indexes a different horizon of spatial and material grammars made within the (im)possibility of Black femininity. Betty's invitation is predicated on the demonic grounds through which (white) property and gender become thinkable. These demonic grounds, which Katherine McKittrick describes as "the absent presence of Black womanhood" or the "spatial unrepresentability of Black femininity" are central to constructing and policing property as a fact of space and a cisgender relation.[48] Central to imposing the racial force of property but cast outside the whiteness of "man" and "woman," Black women are at once contagious (and thus surveilled to regulate the color line) and unthinkable (and/or infinitely monstrous).[49] The proprietary relations of white ownership ensconce Black femininity as impossible if not unnatural, making Black women's fugitive intimacies necessarily a different relationship to matter.

Betty taught me how to see trash as a loophole: "What did they teach us in grade school: Reduce, Reuse, Recycle? Ha, I'm green as shit!" She would wash her shoes with the things that she could swipe: motel shampoo, liquid bathroom soap, windshield wiper fluid. If she could swipe it—"Take a little something for me? Take a bit for my girls? Why not?"—she would take it. "And you know, it's like, Who the fuck gunna do this shit? These people out here being like we got to save the environment, from what? From us?!!" She was gesturing toward the motel, implying a contradiction in what she understood to be the terms of "environmentalism." "I'm out here living a totally different way, running from cops, and the irony is, the trash provide for me more than anything else ever did." Earlier, Betty had shared moments of her family falling on hard times and the things they did to survive. For her, "the trash provide for me" was also a reference to how she felt about local shelters and the way they would discourage her behavior as unhygienic:

People are so afraid to get dirty, to be a little dirty. They hide it from their homes; they take the trash outside like they didn't produce it. Same with hiding it from their wives! They come find and fuck me in an alley somewhere. They hide it and they hide it. But guess what, it all piles up somewhere. Even if you're lying, it's true somewhere. It piles up and piles up and someone's getting fucked! I don't live like that. I live right *here*, and you know what I see when I look in that pigpen? I see my fucking freedom. You living with the trash now, girl.

Betty's wageless living maps waste matter as abundant with the potential for survival. The propertied relations of heteropatriarchy are the wasting relations that, as Betty says, "they hide." Knowing that waste is also abundant—it all piles up somewhere—the girls chart flights through the city to gather the stuff that "gives life." And in so doing, the girls become trashy itinerants, making the discarded and discardable an improvisational mode of life.

As Betty moved, she would sing Missy Elliott's "Work It" like a legend to map the city: "Is it worth it? Let me work it; I put my thing down, flip it and reverse it." Reworking the discards she found on the street, the waste in her hands could be a fugitive commons.[50] The song is not incidental. It's not surprising that as a central figure in the rise of Virginia's Dirty South hip-hop scene, Missy Elliott—born in Portsmouth, Virginia, and known for her futuristic aesthetics layered on top of explicitly sexual lyrics (in songs like

"Work it" and "Get Ur Freak On")—becomes legend for condemned (working) girls, whose stealthy movements remix the Tidewater's geography.[51] In many of Missy Elliott's music videos, dance styles evoke, rework, and critique the cultural forms of slavery through dirty or ratchet aesthetics. For example, enacting the way "surveillance perpetually produces proper classification of bodies," dancers' jerking movements mimic white supremacy's puppetry of the minstrel form in the music video "Work It."[52] The Black ratchet imagination, as Stallings defines it, insists on breaking from the cis-sexist demand to "keep it real" and instead embrace the queer "unreality and performance of failure" that can adhere in hip-hop.[53] As one music reporter put it, "Pharrell, Missy, The Clipse, Timbaland; each of the game's iconoclasts," all from the Virginia Tidewater, "sounded otherworldy, alien, even."[54] What better than the alien-otherworld sounds that emanate from the toxic capture of the elsewhere to be the soundtrack to the criminals' itinerancies that map the pigpen of—or as—freedom?

The girls would also sing Missy when they shared their strategic flights, crafting and recrafting their coordinated movements. Sometimes the time schedules of their maps would change; sometimes their flights would. The maps and their derivations mattered as much as the stability of the key: Remember, be fly. I intentionally invoke a double meaning of "key" here, as both the legend on a map that explains the symbols and the scale around which a piece of music revolves; both meanings signify a way to *read* material relations in time and space. The strategic maps of poor and houseless sex workers wrench open a femme line of sight by reworking the fiction of orderly places and making waste matter an abundant terrain of itinerant fugitives. I present Betty's map here with the caveat that it is always changing. (And if it weren't, I wouldn't share it.) Her mode of living is improvisational, not the proprietary capture of land as property for the fiction of white settlement. Most importantly, the map never stagnates, because like the other girls condemned to the geographies of toxic capture, Betty's improvisations of discarded living mean she's unavailable for complete capture (photographic or textual) and is always still-moving (even if carcerally captured).

BETTY'S KEY: "IS IT WORTH IT? LET ME WORK IT"

Elsewhere, criminalized Black women's cartographic struggles chart new ecologic relations. Because dispossession is a structural relationship to movement, displacement changes what place *is* and what materials and strategies are available to make place *with*. Betty taught me how to avail myself of cartographies that map dirty femme lines of sight. And the first lesson she taught me was that time was a disciplinary regime of property. Starting around 4 a.m., it's time to get off the streets—or at least move the tricks from behind dumpsters; 5 a.m. marks a shift change for sanitation workers, who were now part of Waste Watch, WMI's partnership with police. Explicitly naming how waste and criminality signify on each other, WMI's website states: "Our truck drivers often drive through community streets in the early hours of the morning. This puts them in an ideal position to spot unusual, and potentially dangerous, situations—especially if they are trained to recognize signs of trouble." Now imbued with the police/surveillance power of the state, Waste Watch sanitation workers were authorized to "remove" blight from public space (by which they mean homeless persons and sex workers), identify theft (of waste objects on the street and panhandling), and stop vandalism (including removing graffiti). So by 5 a.m., Betty and the others were gathered in the motel parking lot, singing Missy Elliott, sharing (and revising) new flights through the city, and telling stories of their tricks the night before.

The girls knew the schedule of the city. They knew the tempo of police sweeps and drug busts. They knew when lawyers, doctors, and stockbrokers got out of work. They knew sanitation truck routes and how they differed from private city-scaping companies and to stay away from community cleanup days at the park. They knew that cleaning was a form of policing because it was their personhood and presence considered a contagion threatening the environment. They knew they were trash and they knew how to work it.

Black bodies in a white environment raise the specter of the distinction between humane geographies of the propertied and the dead environments to which Black people are supposed to belong. The Black body becomes an *environmental matter of concern*, suggesting that immanent death, devaluation, and degeneration will follow. This means not only that Black bodies are understood to be *dirt* (matter out of place), but that *cleaning* functions to reinforce Blackness as an excess to the environment. Cleaning is a form of relentless punishment that makes being unpropertied a threat to racial and colonial hygiene. But if, as I have argued in chapters 1 and 2, Blackness is cast outside "nature" to (re)produce the violence of property, Blackness is also cast outside nature to (re)produce property as a cisgender racial arrangement.[55]

It is not only Betty's Blackness but her Black *womanhood*, too, that is an environmental problem. Never appropriately an embodied *owner* (not white and without "class"), Black women on the street are sexually contagious, and their gender is excessive—even excessive to that category "woman." Gender is spatialized impossibly for wageless Black women; the way they inhabit space is always wrong. The way she adorns (or not) her body, her ways of being on/in the street, her movements that violate the colonial fictions of property, are—to put it Blackly—extra. Always too much, her potential contamination makes her an excessive object of Black gendered and material disorder.

But this is precisely the space from which Betty's flyness emerges, from the dirty place of being an object amid other objects. Her aesthetics—oversized, extra, and *fly*—remake the "excess" of Blackness into an eco-ethics that reject the civility of whiteness as environmentally sound. Though her aesthetics—again, oversized, "extra," and fly—might be policed as disorderly object-practices, by adorning herself with "trash," Betty becomes a fugitive ecologist. Betty reworks what it means to be an environmental "problem"; she restages the waste of the city as a place to elaborate a new genre of being, *trashy*. Watched *as waste*—authorizing anyone (but particularly sanitation workers) to render her "captured"—they cannot strip away the modalities of refusing to construct a *Black* womanhood as a propertied relationship between patriarchs or an a priori biocentric fact of the world. Instead, Betty's flyness—her femme trashiness—makes collective identification with matter, aesthetics, and the object-practices of surviving gendered punishment part of Black ecologic thought.[56]

BETTY'S KEY: "I PUT MY THING DOWN,
FLIP IT AND REVERSE IT"

"Now, this is where things get more complicated," Betty explained: In the back of some of the pharmacies that had replaced the girls' homes are clothing donation boxes for the Salvation Army. Early, maybe before going to swipe pills at a coffee shop, or maybe between eight and ten on a Sunday morning, Betty goes "shopping": Stay close enough to the pharmacy to seem like you've just left it. Grab a plastic takeaway bag from the garbage and put a bottle filled with fluid inside to make it seem like you just bought something too, and then, when you see a minivan drive up to the Salvation Army drop off-box, offer to "take it for them, and comment on how adorable their kids are." Now, depending on what you get, clothing can be nice enough to wear. And if not, turn it into cloth that can be used to wrap your head up on those days that, you know, "you just can't get your hair did!"

Also, if nothing fits, you can save it and trade with someone else for pills or even a shower. "We take care of ourselves by taking care of each other."

Being trashy, as Betty articulates it, is forged in the interstices of becoming waste objects; it is a way of improvising with waste and living by reworking, remixing, revising. It is a way to notice the tempo of race, class, and gender relations that shape the spaces of the city. And it is a way to identify the hetero(cis)sexist property relations that can be wrenched open, repurposed, and redistributed for aesthetic, self, and community production. Betty's fugitive itinerary continues: 9–10 a.m. is prime pill-grabbing time: "White women on their way to work or distracted by babies at the coffee shops down here always got some good Xanax or Oxy." Midafternoon, when people are at work, is when you find yourself a shower (but stay invisible): "Gas station sinks, bathrooms at the mall, or, you know, if you've scraped together the money, shower back at the motel." You gotta stay fly!

Betty's flights were counter to how gentrification sediments racial regimes of the so-called gender binary. Her domestic life was in public, and it violated the propertied lines that required her distance to uphold. She wrenched open the object relations sanctioned by white property (the clean and orderly city) and assembled them to improvise paths of movement. Coordinating with other women now: send light-skinned women into the mall to steal perfume and nice panties ("panties are key to grabbing those sleazy white men on their way home from the bank or whatever"); darker women go to the gas station and get things to clean shoes and purses, and if you can manage, swipe hair products (remember, *be fly*).

Being trashy and being fly are not only descriptions of Black women's aesthetics but a material grammar that articulates Black women's conditional presence and a fugitive mode of ecologic thought. Untangling and recirculating the knots of property—the trash bin, the Salvation Army drop-off box, the gas station, the parking lot—Betty refuses an environmental logic that polices her as contagious matter. If cleaning justifies and stabilizes property regimes that require Black and Indigenous disappearance and abjection, self-proclaimed trashiness marks distance from ownership's central categories ("man" "woman," "heterosexual," "white") and critiques an environmental project that obliterates the modalities of the poor and colonized forced to live with the ecological afterlives of colonialization and enslavement, known rather limitedly as climate change or environmental racism. This trashy modality allows Betty to engineer transversal movements not only for herself: she moves lyrically in queer Black fugitive communion.

A SCENE OF ACCUMULATION, OR
A HORIZON AT THE EDGE

So far, New York has appeared only as a fictive place of the "source" of waste production, as if waste is generated by wealth in the North only to choke Black life in the South. But Brooklyn's racial geographies of accumulation by dispossession are the product of toxic capture too. Lead paint in public housing maps onto the neighborhoods hyper-criminalized and patholo-gized as diseased. The diseased are displaced to the surprising amount of Superfund sites that characterize most of the poverty of urban places postindustrialization. Stories of legionnaires waft out across Brooklyn, as rezoning decisions relentlessly remake Blackness not only "excess" but anachronistic to a neighborhood. Gentrification, too, is a violent spatial regime that relies on the mattering of Black life for unceded Native land. Wherever Blackness is constructed as a "hazard," hazardous conditions follow and accumulate. While property's eco-logic might use the matter-ing of Black life specifically and differently across time and place, as I said at the outset of this chapter, Black environmentalisms elsewhere emanate from toxic capture *everywhere*. Emergent in the scenes of waste accumula-tion, and the accumulation of matter in the scenes of Black life, are dirty ecological ethics and new theorists for decolonizing how we live.

It was a hot summer day in 2014 when Marvin whispered under his breath, "Be afraid. Be very afraid." Though he said it too quiet for her to hear, he was speaking to a white woman who, after seeing the "trash" that the five of us were carting—Marvin, Marty/Squee, Terran, Sal, and I—quickly crossed the street and hurried west in the opposite direction. The street that we were walking down, Gates Avenue, was haunted by a history of white fear and Black "degeneration." In the 1960s, the street was a waste-filled artery and it represented, to white people anyway, the "trash" emanating from the hood. Redlining, racial covenants, and the North's de jure Jim Crow had disfigured Black life as if the residents themselves were their own social and environmental problem. In the present, redlining still underpins Bed-Stuy's geographic transformation, often made visible through the juxtaposition of (the racial-sexual dynamics of) a "Black" and dangerous Bed-Stuy with the proposal for a "new" clean and white one. Since the 2000s, Bed-Stuy, Brooklyn, has experienced an influx of white and upwardly mobile resi-dents, skyrocketing rent and real estate prices, and new development from boutiques to luxury apartments. After the city had nicknamed the Bed-Stuy neighborhood Pigsty, a name that came after Bed-Stuy's streets had been ranked as some of the dirtiest in the city, a series of rezoning projects set a new spatial program for the neighborhood in motion.[57] Along with it came the gendered clash of racialized class relations, where scenes of garbage accumulation, like this one, are also scenes of terrified white women clutching their purses as they walk through the hood. It seems a striking coincidence that in 1962, the same street that Black "women and men swept . . . with brooms and used shovels to load dirt and debris into the trailers" in protest of neighborhoods' environmental conditions, is the same street on which Marvin whispers "Be afraid." Gates Avenue once exemplified the acute materialization of Jim Crow's sexist racism, and is now a scene of white women's vulnerability and the fear of houseless Black men's furtive, dirty, and dangerous movement.[58] The color line endures in the production of gender's racial materialization and it is a scene of accumulation—where waste, race, and gender map a story of material relations.

Before I continue to describe this scene of accumulation, let me first describe the crew, who are, among many things, *scavengers*. There was Marvin, whose whispers open this section; he grew up in Bed-Stuy, and though for a time he left and lived outside the neighborhood for a "damn job . . . it didn't feel right, so [he] came back to the hood" in the mid-1990s. Marvin had a kind of chivalry to him that sometimes resulted in my exasperation: "Fine, Marvin, I will walk on the inside of the street!" But it's hard to describe Marvin without describing Marty. Marty, who also goes by Squee (a nickname that I will use interchangeably from here on out), is the tallest in the crew. At six five, Marty towers over the rest and especially Marvin, who is five six. Marty/Squee is extremely gentle, shy even, while Marvin is terse and matter of fact. Marty is joyous, which emanates from his wide gap-tooth grin, and Marvin is (in his own words) a "real feet-firmly-planted-on-the-ground kinda guy!" Marty and Marvin are something like brothers. They take care of each other, though I think Marvin would assert that he takes care of Marty. Marty, on the other hand, would probably say that he lets Marvin take care of him because he needs to feel like a good man. Then there is Terran, who is kind but often doesn't say much. He's not shy, but also not immediately assertive, though when he has to break up an argument between Marty and Marvin, he is. At first I thought Terran was uncomfortable around me, and that adding a woman to the crew had made Terran more reserved, but I came to realize that Terran was not quiet because he was uncomfortable; he was quiet when he felt safe.

Finally, there's Sal, the proprietor of a junk shop for which he had hired these men to scavenge. He had opened the shop for a number of reasons, but one of the most important was "watching so many brothers fall on hard times." The men who worked for Sal had indeed fallen on hard times. Houseless and wageless, these men lived on the streets, most of them displaced by the geographic and economic changes of gentrification. Scavenging was a way to live, a way to make home, self, and gender, with the material arrangements of dispossession. And it was a way to hold on to stories from past eras of Bed-Stuy, something they did to preserve the neighborhood's history before it was "cleaned." Sal was a talker, a storyteller, a people's historian of sorts, and although the dirtiness of the neighborhood was the injunction upon which rezoning and development projects sought to make Bed-Stuy a more "attractive place to live," Sal and his crew taught me about the materiality of change, or perhaps it was something more like the other places in which history could be found.

"Rather than try to convey the routinized violence of slavery and its aftermath," Saidiya Hartman writes, "I have chosen to look elsewhere and consider those scenes in which terror can hardly be discerned."[59] Powerfully indicting the "exacerbation [of an] indifference to suffering" reproduced by staging the scenes of incalculable violences of slavery, Hartman examines "scenes of subjection" to unearth the terror and possibility of self-making.[60] The scene of accumulation is an homage to this method, a desire to avoid exacerbating suffering while simultaneously tracking the forms of subjugation that congeal through the accumulation of scenes such as this. This scene is of our living, and yet it is also a scene that seems to resonate across Black literature, Black biography, and Black art. And I see this accumulation across *genres* of Black telling, depicting, and restaging art in alterity as something that matters. Revising our gaze to see femme lines of flight requires incorporating these genres to showcase how this scene also offers what Ellison calls the "re/appearances of queer femininity" that disorder our categories in order to re/appear a dirty horizon.[61]

We had just come from a partially demolished site, scavenging scrap metal. Some of us were pushing shopping carts, some of us were carrying the siding from old fridges. Marvin had a bag of silverware slung over his shoulder, and it jingled when he walked. And the woman clutched her purse and averted her eyes as if Marvin had jingled *at her*, like a whistle. I can only speculate about what the white woman who panicked and crossed the street was thinking on that hot summer day—*protect myself, danger ahead*. And so instead of focusing on her per se, I want to note that if not for her need to run, this scene might not be a *scene*. Sal looked back at us—Marvin, Terran, Squee, and me—with a guileful smile and said: "To her, we're cock-a-roaches. The thing that these white girls don't know is out here at the edge of *her* world, in the life, we know that cock-a-roaches can survive anything. She's not wrong to be afraid. She can't survive the things we've survived. She just doesn't understand what she's afraid of."

Looking back over my shoulder, I watched her feet quicken. All at once, Marvin mumbled under his breath again; Terran grunted; Sal shook his head. Squee whispered, "They're so scared of us. But why?" Squee was no longer referring to the white woman specifically but to what she represented. He was talking about what it's like to be apprehended as a threat and the way the white gaze turns his height, his hands, his body, and the (waste-filled) environment through which he moves into a form of contamination that can only be contained by criminalization. He is asking us something that all of us know—how that moment of white fear all-too-quickly snaps into focus an accumulation of scenes that constitute the background and foreground of race, class, and gender relations. He is asking us something that we cannot answer because the accumulations are unwieldy and not of our own making. He is asking us a question that in order to answer requires that we engage in misnaming ourselves, and as such, we will have to decide if that misnaming can name something true.

"Because I stink or because there's a pothole?" Squee continued. "I swear, it's like they walk into the hood, take one look around and run away." We all laughed. The elision between our being (even if smelly) and the pothole was illogical. And yet it rang painfully true.

So true in fact that scenes like this often use waste to stage alterity through different genres. From the waste aesthetics of multimedia artist Wangechi Mutu and photographer Fabrice Monteiro, to garbage blowing through Harlem in the opening passages of Ann Petry's novel *The Street*, to the famous tussle with a big Black rat in the opening passages of *Native Son*, Black life and discards seem to get caught up in one another.[62] But who gazes and tells tales about this accumulation matters. It is normally the white gaze that informs our "commonsense" descriptions of danger in the ghetto. After all, it was not Black residents who nicknamed Bed-Stuy "Pigsty." Writing about her childhood in Brooklyn, poet, journalist, activist, architect (and more) June Jordan remembers her father's frustration at the state of Black life: "Look at the damn garbage of Brooklyn. They tink because this a Negro neighborhood, they tink we like garbage, they tink we love garbage, they tink we need the garbage so we can feel at home!"[63] Squee articulated a similar kind of exasperation, "Like I'm the reason that this place is falling down?"

Marvin interjected with frustration: "But we *are* the reason, Squee."

And in his resignation to the materialist logics that subtend anti-Blackness, Marty sighed and said, "I know. 'Cuz we're Black."

Accumulations of refuse congealed where the sidewalk beveled, and trash spilled out over the lip of the municipal garbage bin at the corner. A thin layer of construction dust stuck to our hot skin. Even the breeze rushing through Brooklyn's wind tunnels couldn't lift the dust from our bodies. But it did find us. Just like Anny Petry describes in the opening passages of her novel *The Street*, the wind always finds the discarded.

> There was a cold November wind blowing through 116th Street. It rattled the tops of garbage cans, sucked window shades out through the top of opened windows . . . and it drove most of the people off the street in the block between Seventh and Eighth Avenues. . . .
>
> It found every scrap of paper along the street—theater throwaways, announcements of dances and lodge meetings, the heavy waxed paper that loaves of bread had been wrapped in, the thinner waxed paper that had enclosed sandwiches, old envelopes, newspapers. . . . It even took time to rush into doorways and areaways and find chicken bones and pork-chop bones and pushed them along the curb. . . .
>
> It found all the dirt and dust and grime on the sidewalk and lifted it up so that the dirt got into [pedestrians'] noses. . . . It wrapped newspaper around their feet entangling them until the people cursed deep in their throats, stamped their feet, kicked at the paper.[64]

Chip bags, coffee cups and other discarded matter rolled down the streets of Brooklyn. And in the white imagination, the garbage always finds us, misnaming why it sticks for a reason.

"What is she afraid of?" I snorted rhetorically.

Sal laughed, throwing his head back. "Everything! Nothing at all!" Taking my shoulder, Sal shook my gaze away from the white woman's path. "You know what she's afraid of." Our conversation was happening in angry but hushed tones, and I wanted to tilt this woman's world off its axis. As if reading my mind and with my shoulder gently in his hands, Sal shouted, "We're monsters, haven't you heard?!" His shout scared her, and her quick steps turned into a run. Though Sal had managed to make me laugh, my anger was still at a fever pitch and my eyes were locked on her. *We're monsters, haven't you heard?*

This scene of accumulation is an accumulation of many things: of gender's racial arrangement of matter, of the way dirt and trash signify danger, of the way whiteness's ever vulnerable "purity" requires distance to maintain. But this scene is also a cumulative lesson about the dirty horizon in which gender is an arrangement of the Black, the white, and the trash. When the white gaze changes the substance of survival into a monstrosity, "working through the debris of existence," Cajetan Iheka writes, "also provides an opportunity for healing."[65] Though Iheka is writing specifically about the waste aesthetics of Fabrice Monteiro's series of photographs called *The Prophecy*, the analysis rings true of the discarded lives who make use of, find themselves in proximity to, and wrench open new genres of being with waste. Instead of prioritizing the white disfiguration of Blackness here, I want to prioritize how and why people use their (mis)names to produce counter-presents—a way that people find freedom in the hostile environments that shelter us.[66]

In the horizon of Pigsty/Bed-Stuy, pathology is intertwined with monstrous ways of being. The lines of flight from degraded Black flesh to the degradation of nonhumans and matter offer new forms of relationality emergent in what Simone Browne might call "the Black gaze." Gazing back, though, and from the violence of containment, the accumulation of the discarded makes for different kinds of intimacies. Here, in the elsewhere, where "cleaning" and "clearing" are not so far apart in their violences, the (sub)human and the nonhuman gaze upon each other's disfigurations, productive of new ways to relate.[67] Tina Campt writes in another context, "Tethering blackness to the concept of the gaze is a combination of terms that begs the question of whether a Black gaze is an attempt to invert a structure of visual dominance or domination for the use of the dispossessed. And if this is not the case, at the very least, [a black gaze] raises the question of what, in fact, is its true purchase?"[68] Zakiyyah Iman Jackson's analysis of *Beloved* is informative here. In working through an analysis of Mister, a rooster, gazing at Paul D, a slave, Jackson writes, "*Beloved* identifies the site of a potential breach in the epistemological project of the humanistic perspective." Though Jackson argues that an inability to attend to an animal's perspective is a foreclosure produced by enslavement, the rooster's gaze "invites a critical reopening of the orders of ethical authority and ontological distinction . . . not as a context of investigation but rather as the object to be critically reexamined."[69] In calling us monsters, Sal is doing something, something more than what some readers might worry is just willful degradation.

The stuff of Black living is abundant with alternative relationalities, like the moment in *Beloved* that Jackson identifies. The skewed life chances in the afterlives of slavery make alternatives a necessity. For example, Marvin tells a story about himself. And the fact that he tells it often—across contexts—and that no one interrupts when he tells it, even if he's already told it that day, exemplifies how the story means much more than a single childhood moment. The story, and it's recurrent telling, is critical to how Marvin understands himself. It goes something like this:

When I was young, my best friends were the damn roaches. I was the youngest of six siblings, all girls. By the time my mom had me, she didn't have no more money to buy clothes, so I wore my sister's hand-me-downs. I hated it. The kids at school made fun of me. Always calling me a queer and shit. So I didn't really have friends. I had my roaches. I made like a whole playscape for them you know. I would find junk out on the sidewalk, like I would take some blocks, or a stick, or even just some paper and I would make a little playground for them. I'd share my food when I got home from school. I quite enjoyed their company. One day, I'll never forget, I was wearing a red shirt with a pocket and some kids were bullying me. And one of my friends came right out of my pocket and scared the living shit out of them! Oh, it was the funniest thing. They never messed with me again because I would just say "you're just afraid of a little bug?!?!?!"

We might think about the fact that Marvin tells this story again and again as iterating on Jackson's point about the latent potentiality of a gaze that disrupts Blackness's animalization: "Rather than *a projection of Paul D's trauma*, the gaze of Mister [the rooster]—the exchange of glances between Mister and Paul D—takes on the quality of a caesura, a disruption of the prevailing grammar of gender, knowledge, and being."[70] In Jackson's analysis, this breach is made possible by Toni Morrison's storytelling, textual analysis, and narrative style; we might think similarly about the story Marvin tells of himself—a story he tells over and over because it is formative to his sense of self in that it, too, is disruptive to the reigning ideology of gender, knowledge, and being.

Animalization is part of an anti-Black environmental regime of truth, with specific meanings for urban poverty. Famously in the opening passage of Richard Wright's *Native Son*, Bigger—the titular character around whom the poverty of Chicago's segregated neighborhoods, competing visions of

Black masculinity, anger, violence, and the desire for freedom by any means necessary coalesce—fights and kills a big black rat. After a tussle that takes nearly three pages, Bigger teases his sister with the rat's dead body, angering their mother. In their cramped apartment, Bigger attempts in defiance to get away from her by lighting a cigarette but simply finds himself on the other side of the room looking out the window as his mother says, "We wouldn't have to live in this garbage dump if you had any manhood in you."[71] In an essay titled "How 'Bigger' Was Born," Richard Wright reflects on the birth of Bigger as a lifetime composite (an accumulation, so to speak) of defiant Black men who spanned his childhood in Jackson, Mississippi, and then the kind of men produced by Chicago's physical environment: "noisy, crowded, [and] filled with the sense of power and fulfillment."[72] "There was not just one Bigger," Wright details extensively, "but many of them, more than I could count and more than you suspect."[73] Though Marvin is not a writer, he is, like all of us, a storyteller, whose stories come to form a way to narrate cumulative scenes that define a self. And his narration, not unlike Bigger's tussle, is laden with potential.

For Wright, the context of urban segregation, labor struggles, art, and all forms of stimulation "made the Negro Bigger Thomases react more violently than even in the South." The shared economic but distinct geographic condition of segregation in Northern cities allowed him to "begin to see and understand the environmental factors which made for [Northern Bigger Thomases'] extreme conduct." *Native Son* culminates in multiple scenes of violence; his accidental killing of Mary Dolton, the daughter of a wealthy white family for whom he is a chauffeur, leads to a series of escalating violences, including the rape of his girlfriend Bessie, his arrest, and eventually his state-sanctioned murder. Many critiques and meditations on *Native Son* have usefully opened up a world of Marxist analyses of segregation, literary criticism focused on the stories of Black militancy, and the presage of the Black Panthers' revolutionary politics; however, my interest is in the environmental concern (of gender): a working-class Black man's struggle with a creature with whom he shares a fate, a creature that represents his "loss" of masculinity, which in death foretells Bigger's relationship to violence, materiality, and the gendered race relations of becoming. When Bigger is eventually arrested, he's called "a black ape"; the constraining and caging of Bigger is also a way to show how Bigger *becomes animal* because of his environmental and social conditions.

Marvin's story is not that of Bigger; it is in fact a disruption to the gendered environmental telos of *Native Son* and the often materially determinist readings of Black poverty. Countering the presumption that poverty and economic deprivation turns one into an "animal," Marvin tells a story of friendship, perhaps even kinship with the pestilence of poverty. The roaches help Marvin make time and place within what he feels is a poverty-induced loss of masculinity vis-à-vis his sisters' clothes and being called "queer." Ironically, his expression of masculinity is not in the dominance over "pests" but in the way roaches become co-conspirators. Though Marvin's story is undoubtedly one of manhood, it's also a different vision: a Black revision on the white gaze of poverty. In the dirty horizon from which Marvin gazes at the world, his masculinity is not one based solely on the historical injury of once being property.[74] Instead, Marvin's tale of becoming a man is something more like *becoming a pest*.

The monstrosities that lurk in Blackened environments challenge the gendered relation to ownership and property as the only way to *be* in the world. In the scenes of discarded accumulation, likening oneself to pestilence or waste is not a willful act of self-degradation; it is an epistemological and material reality with which poor, homeless, racialized, and disabled people live historically and relentlessly through time. Though the white gaze aligns itself—illogically, coercively, violently—with a presumed imagination of civilization's genders ("man," "woman"), this gaze is constrained by property, foreclosing (and narrowing) the imaginations of relationality.

Like Jackson says of the "literary and visual artistic genres and traditions" of Wangechi Mutu and Audre Lorde in the closing passages of *Becoming Human*, Marvin and Sal, too, narrate Black living as relentlessly abundant, "performing innovative philosophizing and contrary aesthetics using what Saidiya Hartman has once called 'the visceral material of history.'"[75] Here those materials are the degraded matter of the hood: trash, roaches, Black flesh. And in *living*, Marvin and Sal mount an ecological revision that Jackson, continuing her discourse with Hartman, argues, "'breaks forms, breaking them open so other kinds of stories are yielded' and other philosophies of being can be felt and known." What if there are other things to be than human? What if the continued relations of subjugation, geographies of containment, and forms of policing that render Black masculinity a danger to white women are also a material terrain from which to rewrite the meaning of matter and being in *this* material world? Perhaps the presumed animality of subhuman "Black(ened) flesh," to use Jackson's term, and the presumed unthinking nonhuman relate differently elsewhere—where there are other things to *revere*—a place from which to conspire an alternative set of relations that revise property's grammatical hold—including its hold on gender.

"Lord, this place has been through some *shit!*" Marvin said to me one sunny end-of-summer day. We were about to break into a building that had been seized by eminent domain. "Let me tell you what, we've always been expected to live with the trash. We are the trash. No difference." He was jimmying open the lock around the city scaffolding. I had watched him do it countless times before. Construction had not yet started, but the scaffolding and the Department of Building signs were enough information to know that development was imminent. While he was talking, the wind kicked up the trash on the street—chip bags, a loose sock, rubber bands, and some cellophane, which wrapped itself around his foot as he kicked the ad hoc door and popped open the lock. He threw his hand up in exasperation while we talked, "People coming in now act like we're dangerous? These new shops and whatnot, what are they even for?" Marvin pointed to a new coffee shop with macramé in the window. "Like, what the fuck even is that shit? Looks like a goddamn spiderweb to me! Yeah, they act like *we're* the ones who are dangerous? Maybe we are!" Following Marvin's gaze back out onto Bed-Stuy and the way it was being recomposed, for a moment the trash, the dirt, the Pigsty felt like freedom. *Maybe we are dangerous?!* "You know what? At the edge is a different way of life!"

THE TANGLE OF PATHOLOGY

The crew described in the previous section are criminals. And though some forms of reclaiming recyclable materials are legal (namely plastic bottles), they are highly regulated. For example, in New York City, the scavenging of recyclables is restricted by the use of vehicles.[76] In 2016, the New York City Department of Sanitation (DSNY) cracked down on the use of shopping carts as a violation, resulting in the arrest of many scavengers. DSNY's monopoly over its "commodity" affirms capitalism's wasting relations as natural. By turning the commodity into part of the city's market nature, the DSNY under the mayor's orders also reframes scavenging as "theft"—criminalizing wageless and houseless survival as a way of life. Refusing wasting relations as rational or natural casts Black life as a "hazard" to the nature of property's value(s). These refusals can also reveal the fugitive affinities and degraded femme possibilities of rewriting the tangled ideological trap that white supremacy produces: waste and Black pathology.

Ensnared by the web of racist hetero(cis)sexist ideologies, Bed-Stuy, Brooklyn, has been the subject of many forms of containment, including the containment of garbage. In order to chart our way through the neighborhood along a femme line of flight, we have to, as Betty might say, "get down and dirty." Arcing closer to the dirt requires arcing closer to the tangle of pathologies that bind Black life "elsewhere," and it requires arcing closer to the way pathology makes Blackness a place for waste. "Waste is considered a sign of backwardness and is one rationale for the colonial mission of developing a civil culture of cleanliness and orderliness."[77] The "rhetoric of difference" that determines what is civil and what is pathological "is mapped onto the body of others through a spectrum of dirt-related words."[78]

But it's not just the body onto which these words are mapped; it is also matter and, more importantly, the relationship between the two. The materials that make up an environment, how people use objects, how people are objectified, and how the state accounts for objects all make a difference. As I have said elsewhere, waste is not a metaphor for racist dispossession; it is, in fact, a material reality of racial objectification knotted together by hetero(cis)sexism and classism.[79] But in arcing toward the dirt, of course, there is a danger that we, too, might (mis)name this tangle, reproducing the pathological web in which Black people perpetually find themselves ensnared. This section revises the well-worn story—so worn that it deserves to be torn!—that Daniel Patrick Moynihan once called the "tangle of pathology." By focusing on racial entanglement with accumulated risks, I reanalyze a garbage protest that took place in Bed-Stuy in 1962. In so doing, I propose that we ask what using garbage to *protest* might tell us about the fecundity of dirt. Brian Purnell's work, which rewrites the assumption that Jim Crow "missed" Northern cities, is critical to excavating this tangle of matter and flesh.[80] I draw heavily on his analysis and the interviews he made available to retell the spatial relations that define the parameters of suffocation in 1960s Bed-Stuy. To his analysis, I bring attention to gender's subsumption in the districting of waste collection and the way racist visions of the pathological Black family map degraded matter onto degraded flesh. My goal is to reveal a different horizon, a Black horizon of environmental politics that revises where our environmental stories start.

In 1962, a Bed-Stuy resident-led caravan of cars approached Borough Hall in downtown Brooklyn. "Armed with brooms and receptacles," the *New York Times* reported, "indignant residents of Bedford-Stuyvesant district of Brooklyn staged 'Operation Clean Sweep' . . . to protest what they called discriminatory practices by the Department of Sanitation."[81] As the fifty or more men, women, and children who had staged a neighborhood cleanup along Gates Avenue from 10 a.m. to 2 p.m. reached Borough Hall's steps, they brought with them an accumulation of discarded matter that had been lingering on their neighborhood streets: "old mattresses, a rusty ice box, rugs and bedsprings," and more.[82] The residents of Bed-Stuy were protesting the DSNY's trash pickup schedule, arguing that "garbage pick-ups were made by the city five days a week in white neighborhoods, but only three times a week" in theirs.[83] With the help of the Congress on Racial Equality (CORE), residents distributed leaflets and held signs that said "Operation Clean Sweep" and 'Taxation without Sanitation Is Tyranny." Though the protest concluded the way many Black protests often do—the chairman of CORE's Brooklyn chapter was served with a summons for a traffic violation and his wife Marjorie with a "summons for littering the Borough Hall steps"—residents had grown tired of appealing to the DSNY for relief and so chose instead to deface the sanctity of the law.[84] With trash.

While the New York Times reported on the "Brooklyn Group" as indignant residents who "Flaunt Debris," Black residents insisted that the conditions of their lives were an acute manifestation of racism. And it required revision. If the DSNY and the New York Times took Bed-Stuy's streets to be evidence of Black pathology, Black residents took their streets to be evidence of the environmental truths of racism. Gathering up the materializing risk that concentrated on the street, they remapped the relationship between flesh and matter. As such, they rewrote the truth: Trash is a problem produced by white *orders*; its concentration in Blackened places needs to be redistributed. *Imagine if all the garbage produced in wealthy neighborhoods were dumped on their steps!* Dumping garbage on the steps of Brooklyn's (original) city hall, residents redistributed the environmental risks that concentrate in their lives and staged a retelling of Bed-Stuy not as waste but as a place and a people who refused to be wasted.

Franklin Roosevelt's 1933 Home Owners' Loan Corporation Act was critical to the concentration of waste in Bed-Stuy. The act, which sought to stimulate a fledgling economy during the Great Depression, introduced Home Owners' Loan Corporation (HOLC) security maps to help refinance mortgages that were in default and stimulate homeownership in the 1930s. However, the security maps, which were used by banks to assess the risk of loaning money to potential homeowners, turned race—namely Blackness—into a risky investment. Neighborhoods with Black populations, like Bed-Stuy, were "redlined"—that is, turned into the riskiest investment—making the places where Black people lived places that the city did not value.[85] It didn't help that congealing around Black people's lives was the emergent theory that what was wrong with them was their "underdeveloped" family structure, whose roots lay in the history of slavery. Explanations for the reason why Black people lagged behind economically quickly converted poverty into a problem of the Black family. Based on the fears that Black degeneracy was contaminating to property values, racial covenants—agreements that prohibited the leasing or sale of property to Black people—worked in tandem with redlining to narrow where Black people could live.[86] And Bed-Stuy was one such place.

At the same time, Bed-Stuy was growing. In the 1930s, Bed-Stuy's Black population doubled; this was in part because Harlem's housing stock was shrinking and just a decade earlier, Bed-Stuy was one of the few places where African Americans and West Indians could own homes. In the 1920s, Bed-Stuy was a surprisingly racially mixed placed, with Irish and Italian immigrants and African Americans and West Indian homeowners living side by side.[87] But the large influx of Black Southerners in search of industrial jobs to places like New York, Chicago, Philadelphia, and urban centers across the South had been met with racial violence. Increased lynchings and police violence culminated in the Red Summer of 1919. The national tensions that had been growing around the legalization of segregation in 1896 had produced a violent spatial regime that sought to place the risk of Black people elsewhere to white life.

By the 1930s, Bed-Stuy's racial violence had been codified by the redlining that disfigured Blackness as degenerate and a "risk" to property. And while HOLC security maps undoubtedly tell a story of environmental racism today, they also tell the story of Black women held captive by "the tangle of pathology."[88] The contradictions through which Black women come to *matter* are located in how "objectification is coupled with black humanity/personhood."[89] Writing with the work of Hortense Spillers in mind, Uri McMillan reminds us that the distinction between "body" and "flesh" for Black people is tied to enslavement: "The former is the apotheosis of a liberated subject-position, while the latter is a total objectification, an 'absence *from* a subject position,' a forceful reduction of the body 'to a thing, becoming *being for* the captor.'"[90] Captive flesh was both a critical production of the plantation owner's property and a classification of Black women, whose value adhered in their capacity to produce more property. Put differently, the Black woman's thingification is critical to how she is used as "a sign" of degeneration, if not a plane upon which to define a tangle of risk: the risk of capture was put upon her head at the same time that she was asked to bear the weight of passing on "risk"—*partus sequitur ventrem*, the sign of the slave emanates from the mother.[91]

But the sign that emanates from the Black mother in the 1960s is not enslavement but degeneracy, raising the question, Where does she belong? In *Demonic Grounds*, Katherine McKittrick writes, "The classification of black femininity was therefore also a process of *placing* her within the broader system of servitude—as an inhuman racial-sexual worker, as an objectified body, as a site through which sex, violence and reproduction can be imagined and enacted, and as a captive human."[92] Placing Black women is no easy task, because as McKittrick also argues, it is through the obliteration of Black women, their geographies and geographic knowledge, that geography comes to constitute its very lines of codification. Keeping in mind the way that racial-sexual condemnation is critical to the production of Black geographies, I want to think of "degeneration" as a line of flight that allows us to see how degraded matter maps degraded female flesh. I do this to look at the way the "tangle of pathology" is not, as Moynihan would've put it, what Black youth are caught up in but rather how hetero(-cis)sexist racial capitalism produces a threat of its own making.

Though Moynihan had not yet put out his infamous report that saw the "matriarchal structure" of Black families as the source of Black people's failure to succeed, the demonization of the Black family was already well established. Writing about historian Philip A. Bruce's (1889) *Plantation Negro as a Freeman*, Dorothy Roberts points out, "Bruce explicitly tied Black women's sexual impurity to their dangerous mothering. He reasoned that Black women's promiscuity not only provoked Black men to rape white women but also led the entire Black family into depravity."[93] It should not be a surprise that this long-held belief about Black women's depravity and their hypersexual nature, which leads her to "procreate with abandon," evidences itself on residential security maps as "infiltration."[94] As Mel Chen has argued, the ways we animate our racial and gender hierarchies through language "construct deflated animacies for some humans" while racializing others.[95]

Arguably, this was also extended to buildings: "Obsolescence and poor upkeep" are sited as detrimental influences on the same line as "Infiltration of Negroes," leading to the map's clarifying remarks, which state, "Colored infiltration a definitely adverse influence on neighborhood desirability."[96] The collision of "infiltration" and "degeneracy" made for a simple mandate: to contain a growing Black population based on the assumption that Black women's degeneracy was contagious. But these conditions of containment, which required the commonsense assumption that the Black family, and Black women in particular, were to blame, also produced an unwieldy accumulation of garbage on the streets of Bedford-Stuyvesant. Redlining not only operated on the degradation of Black womanhood, or rather the Black family as a problem for which she was at fault, but also tightly wove her to the degradation of matter, the place to which "the trash" *belongs*.

The spatial districting of waste pickup evinces the tangle of degraded flesh and degraded matter. In order to see its operation, we need to look at how the DSNY managed garbage in Bed-Stuy and the way waste districts are constructed. First and foremost, waste districts are constructed by number of domiciles, a unit of measure that was, on its face, not about race.[97] DSNY districts assume that the number of homes is a proxy for the number of residents, a number itself based on the white heteropatriarchal fantasy of a two-parent household. Population density determined how DSNY districts were mapped, meaning how often the garbage was picked up by municipal workers. But density was measured by *type* of domicile, not the *number* of people. Bed-Stuy's population, as previously mentioned, was growing, and it was growing in the white commonsense imagination as racial infiltration. Historian Brian Purnell writes, "Real estate agents played on racial fears and plummeting real estate prices to convince white homeowners to sell their property. The area's brownstone and limestone houses, became carved-up into three, sometimes four apartments. On top of that, bigots refused to rent apartments or sell homes to black families in other parts of Brooklyn, which would have relieved overcrowding in the neighborhood and placed less strain on its housing stock."[98] The increasing Black population and decreasing housing stock meant that more people were living in houses than the DSNY districts were counting. The on-its-face "race-less" measure of domicile units was actually a *racist* justification for continued neglect, which itself actively produced the degradation of Black life.

If the type of domicile mapped DOS frequency of neighborhood garbage collection, then by the 1960s Bed-Stuy was suffocating under the weight of its own waste. To add "insult to odiferous injury," to use Brian Purnell's quip on the racist history that led to Operation Clean Sweep, not only did DSNY workers not pick up the trash that was overflowing street side, but they'd often leave residents' trash cans full. According to Brooklyn CORE member Maurice Fredricks, "The garbage men would come by, supposedly to collect the garbage, but when they left, the place would be very filthy."[99] Another member echoes these complaints about spilled trash in the streets:

"They didn't clean it up. They left it there. And that's what was so disturbing about them taking up garbage in black neighborhoods. . . . In other neighborhoods, if they spilled some garbage on the sidewalk or in the street, because they have brooms and shovels on the side of the truck, . . . they didn't leave it lying in the street."[100]

The complaints leveraged against the DSNY echoed the same concern: DSNY collection protocols were discriminatory. One resident went so far as to document how much trash was being left on the streets, tallying the number of trash cans emptied, litter on the streets, and bags left out before and after pickup. Still, no amount of "evidence" could penetrate the DSNY's racist refusal to take Bed-Stuy's trash problem as a problem of collection. DSNY's lack of trash collection seemed to treat Bed-Stuy like trash was just "part of the neighborhood."[101]

Arguing against CORE and neighborhood activists, sanitation commissioner Frank J. Lucia argued that frequency of collection was based on population density and *not* the neighborhood's race and class composition. The denial of racist neglect, however, smuggled in gendered visions of "the home." The materialization of racist-sexist spatial relations took the form of trash on the streets, but "trashy" streets are quickly turned into the heteropatriarchal "failings" of Black women. Though the sanitation commissioner did not explicitly cite women as the source of disorder, the 1965 Moynihan Report, published just a few years later, insisted that the disorders of Blackness spread from the degenerative Black home to the street. As such, the failure of Black-women-led households was the failure of domestication, and the lack of economic progress was linked to pathological behavior. Within this racist-sexist imagination of disorder, Black behavior and the street were interchangeable: Pathological behavior made "bad" environments, and the environment made "bad" behavior. Thus, the assumption that Black living wastes space, or wasted space is the result of pathological Black living, is also linked by a *sexist* determinism that (mis) reads the materiality of racism.

The conflation and equation of the condition of "the street" with Black residents allowed the DSNY to pose "solutions" to Bed-Stuy's trash problem that ignored the *production* of Bed-Stuy as a dirty place. According to the commissioner, "Bedford-Stuyvesant's garbage problem . . . result[s] from its inhabitants' behavior, not racism or the city's negligence."[102] In response to the mounting pressure from Brooklyn CORE activists and borough officials who demanded change, the sanitation commissioner responded with two "color-blind" promises: more anti-litter enforcement and a commitment to work with civic groups on waste education.[103] Rejecting allegations that DSNY collection practices were racist, Lucia insisted that five-day collection in Bed-Stuy was just not possible without changes to the department's budget. Not unlike contemporary responses to the politics of blight, Lucia doubled down on waste *surveillance* and civic *education*—in other words, the policing of Black people's behaviors.[104]

"Solutions" to the pathologies of Blackness usually entail increased policing.[105] Signs, summons, and public health officials were used to surveil the garbage in the neighborhood. The department officials Lucia was willing to spare were sent to reinforce litter summons and levy fines against landlords and residents who "neglected" their neighborhood. This additional supervision into the neighborhood, he claimed, "would 'make possible increased control over the general situation.'"[106] Throughout the neighborhood, DSNY placed 75 new "No Dumping" signs and an additional 237 trash bins. The 1,124 summonses for Health Code infractions that were issued only served to proliferate "evidence" that the city was not ignoring the problem (as the community had initially charged) but was "empirically" documenting a cultural problem: poor people are dirty, poor in mind, and need to be educated.[107]

But Black life is what makes Black studies possible, and residents had studied their streets.[108] They memorized trash routes and cataloged trash accumulations using, as Black people always have, interdisciplinary methods to analyze the world.[109] In so doing, their protest would make *trashing* a critique, and this trashing would deface the pedagogy of property's vested interest in white superiority. Brian Purnell writes, "Operation Clean Sweep had a strong effect on demonstration participants and local politicians" even if it "could hardly bring about the type of changes Brooklyn CORE demanded."[110] Purnell notes, "Some participants, like Robert Law, were empowered by the demonstration. 'Throwing that garbage was emotionally gratifying. It was like, here, take this garbage back!'"[111] This emotional refutation of the city's discrimination is a powerful reminder that the degradation of matter and people is an imposition. Garbage is political and it's produced by racial capitalism. Discarded Black life makes use of its political potential to revise the status quo of property's eco-logics. Making waste proximate to Black life is also the making of a political terrain of contestation. The reigning property relations that produce and then contain Black life *as if* it were a degenerate infiltration of pathological behavior also produce a wayward politics of unruly potential. My investment in thinking through the production of degraded matter and degraded flesh here does not seek to undercut what the protest was about—the state-sanctioned racism that shaped and still shapes Black life. Instead, I want to think about the lines of flight that emerge from such a disfigured production of state violence and the latent potential of discarded people to use the matter and (mis)names that degrade to tell a different story: the environmental story of Black life.

TRASHY AS FECUND

The fecundity of trash begets wayward kinship—sometimes with pests, sometimes with matter. It is a way of thinking about the imminent abundance of "the space[s] created by deviant discourse and practice" in which, Cathy Cohen argues, "a new radical politics of deviance might emerge."[112]

Trash is fecund precisely because it is tied together by the tangle of pathology, the place of deviance in which other kinds of politics emerge. We might even think, like Betty does, of trash as a place from which to refute the grammars of property. If racial capitalism is always sowing the seeds of its own undoing, might not trash be part of that undoing? And if so, is it not a weapon in the hands of those whose lives are lived under its accumulated weight, those whose lives are made monstrous through the production of waste, risk, and violence? Is trash and its fecund imaginations a condition of possibility for refusing the codes of property that make an uneven world? In *The Undercommons*, Stefano Harney and Fred Moten suggest that there is an ethics of being amid dispossession that doesn't guide people toward property as a way of life but points toward other horizons and genres of being. Within these ethics, trash is part of an "ecological account of resistance," an environment that white men dare not enter and white women, in their raced and gendered subject constitution, are disciplined to fear.[113] Trashy landscapes, as Betty, Marvin, Sal, CORE members, and Brooklyn residents of the 1960s reveal, demand and enable other tactics, ways of being above and below the eyeline of propertied men and their women alike. Perhaps what is so fecund here are the ways that trash could contribute to a revolution itself if we learned to read it as both a product of violence and a weapon forged *against* that violence.

The fecundity of the dirty imagination requires opening up what "trash" means at the threshold of capture. Trash becomes matter made alongside the fungibility of Blackness, a fugitive site of knowledge production through which alternative grammars of seeing and being emerge. In "The

Rain (Supa Dupa Fly)" Missy Elliott debuted her now-iconic futuristic aesthetic in a Black blown-out suit. Close to the camera we look upon Missy's trash-bag aesthetic through a fishbowl lens. Cutting back and forth between her futuristic trash aesthetics and light monochrome joggers, she drives across the postindustrial landscape of Portsmouth in a Hummer. (Profane!) Kicking up sand and dirt along the marshy roads of the Tidewater's toxic capture, she raps, "I'm driving to the beach / top down, loud sounds, see my peeps." I can almost imagine Betty there, dancing with Missy on the beach. In the gray misty day of the music video, Missy in (nearly) all white is complete with chains and studded white gloves that she flashes in the low-down camera, revealing her waves gelled to perfection. What Missy presents here is an aesthetics counter to (white) visions of ecology. It is not, however, counter to Dirty South aesthetics that continuously upend pure, quiet, and pristine environments that never show signs of violence or capital. It is precisely at the juncture of "wasted landscape" (the marshy Tidewater) and "wasteful things" (the Hummer) where the dirty imagination proliferates new genres, new readings, and new refractions.

DISPOSAL

As David Pellow writes at the outset of *Garbage Wars*, "Solid waste is a fact of life."[1] And it is seductive to see consumption and waste generation as a material fact that we all share. For example, the EPA's thirty-year study on waste generation and disposal excludes income as a predictive factor of waste generation, producing the highly sited statistic that Americans generate 4.38 pounds of waste per person, per day.[2] But in Pellow's history of race, waste, and labor in Chicago, the environmental historian insists that "those social groups that consume the most natural resources (environmental 'goods') and create most of the waste and pollution are the least likely to have to live or work near the facilities that manage those environmental 'bads.'"[3] In other words, disposable income correlates with disposal in two ways: the more disposable income, the more waste generated, and the higher the percentage of disposable income, the farther one tends to be from the sites of disposal.

WASTE GENERATION

There are many ways to tell stories about waste generation: through the changing technologies of manufacturing and population booms associated with industrialization; through changing consumption practices over time; through the changing practices of collection and recovery, such as ragpickers, gleaners, peddlers, and "the domestic arts," as well as garbage hauling and the class wars waged over it; and, archaeologically, through the social ideologies, including stories we tell ourselves that are often radically different from garbage as a fact of material cultures.[4] These analyses allow us to approach waste—trash, refuse, garbage, rubbish—in a variety of ways, including by distinguishing the language through which we render these objects know-able and learnable.

By now, readers might very well be frustrated with my use of *waste* as I slip back and forth between *waste, trash*, and *toxicity*, without justifying why I use which word when. But I am reminded of Dominique Laporte's playful theorizing that Georges Bataille's *Accursed Share* might've anticipated. As summarized by Gidwani and Maringanti in their work on the waste-value dialectic, Bataille noticed that "restrictive economies of *meaning* that seek to dominate can-not contain the sheer excess" of meaning and matter within capitalist society and that, most importantly, such "pool[s] of excess . . . cannot be *predictably annexed* for its growth."[5] Try as the EPA or (waste) scholars might to generate stable meaning, sometimes slippery metaphors are in fact part of society's management of disposal. "Surely, the State is the Sewer," Laporte writes, "not just because it spews divine law from its ravenous mouth, but because it reigns as the law of cleanliness above its sewers."[6] Laporte's critique of the state is similar to the work of archaeologists William Rathje and Cullen Murphy in *Rubbish! The Archaeology of Garbage*: "Garbage has often served as a kind of tattle-tale, setting the record straight."[7] If Laporte's quipping of the state-as-sewer is, as I read it, a use of language (excess meaning) that better sets the record straight, we can see how *distance* from disposal often aligns itself with the power of the modern state and its attendant institutions that police and clean. But what, then, characterizes proximity to disposal, and how might it revise our stories of waste generation?

PROXIMITY TO DISPOSAL

The 1987 report on *Toxic Wastes and Race in the United States* was the first national study to show that race was a central factor in the siting of hazard-ous material. One of the report's central arguments was that under Ronald Reagan's program to further distinguish federal and state powers, who was responsible for the environment, and particularly the management of haz-ardous waste, became ever more incoherent. The report points to labor sec-retary William Brock's decision that sanitation and clean drinking water for farmworkers was *not* the Labor Department's responsibility as a precedent that absolved the federal government from intervening in the toxification of people of color's environments, particularly *at work*. Further, the report ar-gued that in the era of Reaganomics, presuming "action by states would be preferable and more effective" than action by the federal government would end up benefiting polluting industries who had the power to lobby states with less environmental regulations.[8]

And this is precisely what happened.

In a prescient analysis, the report warned that increasing the responsibility of state agencies "without corresponding increases in resources" would result in increasing risks to Black, Latinx, Pacific Islander, and Indigenous peoples who were already more likely to live in proximity to hazardous dumps or work with hazardous materials.[9] Understanding how the state acts in the interests of capital, the report accurately diagnosed that when regulators deal with pollution, they are motivated by economic efficiency, which only serves to intensify racial discrimination. Thus, the expansion of the EPA's regulatory purview to include waste management simply placed more responsibility on state agencies experiencing budget reductions. Lower land values produced by racialized disinvestment and white flight, deindustrialization contributing to the plummeting value of farmland, and the privatization of the waste industry as a salve for the gap between the state and the federal responsibility to manage the environment all converged on the South, which became the nation's prime site of disposal.

If this racial-spatial story of capitalism's management of the environment and its production of excess tattles on the state's approach to disposal, then our stories of waste *generation* require serious revision. As I said at the outset, the statistics are seductive, but they tell a story of individual consumption and offer the illusory promise that individual responsibility will make a collective difference. However, consuming "environmentally friendly" options—such as biodegradable plastic that cannot make good on its non-toxic promises—requires more disposable income, making poverty and its racialization seemingly a site of environmentally problematic consumption. Disposal and the production of disposable objects, life, nonhumans, and land are nefariously yoked together through the death-dealing paradoxes of neoliberalism.[10] The focus on consumption and waste generation is a neoliberal misdiagnosis, quietly skirting the racial-spatial economy that subsumes but cannot annex all that it produces.

BLACK TRASH

In his short but incisive critique of the "raceless" social contract, Charles Mills writes that "the black body would prove most difficult for the macro-body of the white polis to absorb."[11] The "doctrine of discovery" that labored across

Indigenous bodies to "clear the land" so that it might be transformed had also produced an unpredictable object: the slave, who, as property, belonged to a white master but whose personhood and agency was an excess to be constrained.[12] After the Civil War, Mills notes, distance from "the black body" was a necessary part of reconstituting the white body politic. And while the moniker "white trash" marks the admonishment of poor whites who fail to "live up to the responsibilities of whiteness," "blacks *themselves* [became] the environmental problem," making *Black* and *trash* redundant.[13]

The way Blackness is made to (not) matter, or rather made *into* matter, reminds us that the past is not so settled. As Robert Bullard forcefully asserts, "a colonial mentality" that "emerged from the [South's] marriage to slavery and the plantation system" turned the South into a "sacrifice zone."[14] Further articulating the continued utility of racism to the protection of white life as the reigning environmental paradigm, he asserts that "a form of illegal 'exaction' forces people of color to pay the costs of environmental benefits for the public at large."[15] As I suggest in "Infrastructure," waste management and landfills teach us to see waste as something to be disappeared. And disappearance (across distance) makes the relationship between value, Blackness, and elsewhere economically efficient for the management of waste. The elsewheres of risk that accrue benefits to white property are part of the ecological afterlife of slavery, and the legacy of slavery is located in the patchwork of authority, policy, and regulation that "protects" the environment, including in the management of disposing of *the disposable*. The mattering of Black life is critical to waste generation, because the economic relations that produce waste turn Blackness into the disposal capacity for the nation.

4 Black Refractions

Story is the practice of Black life.
— MCKITTRICK, *DEAR SCIENCE AND OTHER STORIES*

Sal's junk shop was a love letter to Brooklyn, an archive of Black history. Sal and the houseless men who worked for him, like Marvin, Marty (Squee), and Terran, were archivists of the people, pedagogues of the oppressed with object lessons to share. If you were willing or wanting, Sal would tell you a story. He could tell a story so engaging that you'd feel as if you were embroiled in the tensions between the Dutch and the British in colonial Brooklyn or were on the block the day that Biggie died. All these stories were a kind of fiction, stories in exchange for capital. In Sal's words, "These little gentrifies want things to *mean* something. They want to believe they've found a piece of history, or maybe they want to be the authors of it. Alls I really know is I sell them their *own* imagination." But these stories were also fabulative, ways of making place and time against the inscriptive violence of real estate and gentrification, imagining what might've been said, who could've been lost, and to make meaning within the impossible condition of being excess, rendered a waste by capital. Redundant.

Saidiya Hartman teaches us that foregrounding stolen voices requires "straining against the limits of the archive to write a cultural history."[1] As a tool of colonial capture, the archives of slavery are a complex place of hegemonic inscription—a place of dispossession tallied in the numbers and ledgers cataloging the sale of people, as much as a place shaped by the logics of possession wherein those same ledgers and sales continue to be proof of enslavement. Sal's shop was such a place, not an archive of slavery exactly but proof of its afterlife. Sal collected the evidence that violence makes *waste* and, through story, spoke about the ways waste conditions (but does not determine) Black life. As Hartman says of the rigors of fabulation, writing history involves "enacting the impossibility of representing the lives of the captives [the dispossessed] precisely through the process of narration."[2] Thus, through this chapter, we work alongside Sal's narrative injunction: "I'm holding Black history in my hands."

To be clear, though, Sal and the men who worked for him were hustlers; the shop was not a fantastical place outside capitalism but diagnostic of it.[3] Well versed in the ways that capitalism racializes the "surplus," they sought to redirect what they could to the hood. And to their own pockets. The lines of violence drawn in the earth by settler-colonial capital's migrations also produce the interstitial corridors where dispossessed people make place and time in the elsewheres laid to waste.[4] Perhaps another space born of the "dirty imagination," the fugitive plans of the hustle index the improvisational rigor required to live *elsewhere*, a mode of living forged under the conditions of toxic capture.[5] Improvising within the traces of capital, however, requires different kinds of reading practices, ones that are responsive to the values of property but not wholly dependent on them either. It requires a reading practice that runs counter to the way property sees Blackened environments as unruly, chaotic, unecological, and, perhaps most importantly, unstoried.

In the toxic capture of a Black backyard (if one has a backyard to begin with), history is hard to find. As I've shown, the criminalization of Blackness is a land-use strategy for managing the environment and justifying the mundane toxification that sustains white life. The un-ecologic (an anti-Black eco-logic of property) makes Blackness quintessentially disposable, criminalizable, and disfigurable; suffocated with waste or managed as a waste of space, unwaged and unpropertied Black living is subject to expulsion. The consequence is not only that houseless urban life is cast as the excess to land (and condemned to slow and quick deaths) but that the mattering of Black life makes the houseless suspect narrators of ecology and untrustworthy historians who, "in the predominant view . . . waste time telling 'tall tales.'"[6] But as Nadja Eisenberg-Guyot argues in another urban context, the tall tales of those thrown away by society (Black drug users) are in and of themselves subaltern histories and political pedagogy.[7]

As Sal would say, "The ghetto is a gold mine," a refrain he marshaled in the face of the rezoning of North Bed-Stuy in 2007 and then South Bed-Stuy in 2012.[8] In response to the neighborhood's nickname, Pigsty, rezoning decisions were aligned with the imperative to "improve" the neighborhood, a place behind time or without history.[9] Unlike the predominantly Black middle-class homeowners that made up Brooklyn's Community Board 3, Sal and his crew of working-class and houseless Black men had different investments in the neighborhood and in preserving its stories. Though responsive to the gentrification underway, Sal and the others' labor was something akin to finding Black history. Marshaling a set of descriptions/stories athwart *Pigsty* is epistemological work that flips *trash/junk/ghetto* into gold. It is social labor that tells stories other than the civilizing teleology from worse to better.[10] Telling stories of people no longer there, the stuff of living, makes the mattering of their individual and collective Black lives matter.

This crew of hustlers, not unlike the (working) girls in the South, have been dispossessed by violent spatialities. Some of them will occasionally say they're from Virginia, and I like to imagine that at some point we were crossing paths in a Black space-time continuum. As it was for the girls in their fugitive living room, the "facts" of people's lives were always up for revision. Condemnation makes waste a condition of racialization, but racialized people make waste into an abundance from which to retell their own stories. Though stories might not intervene in the chronic production of toxic capture or the anti-Black "use"/theft of native land, the stories the dispossessed tell and the material relations they engender make what June Jordan might call *living* room.[11]

As the only woman in a crew of men, I was particularly attuned to the ways women were or were not present, and to how masculinity seemed to be a constitutive element of the fabulative work of collecting and constructing Black life among and through objects. In this chapter, I am attuned to the gendered genres of telling stories about Black history and matter. Thus, I proceed with questions and speculations about the fictive role Black women played in this speculative waste terrain for the men whose labor and storytelling I narrate. Attentive to the genres of "telling" that emerge within discarded Black life, this chapter focuses on how Black women become a genre unto themselves, tales of a lost Black city. If part of what delimits Sal's archive is the history that gentrifiers want, then the fabulative is a way to mark Black stories that can't be possessed, can't be policed, and can't be stolen, even when sold. Sometimes elaborate, sometimes mundane, quiet, and solemn, these stories are rigorous attempts to name that which circumscribes Black life and makes space for the way Black living exceeds circumscription.

I should note that there are no pictures; I rarely took them while doing fieldwork. Sal and his crew work in tension with the forces that strip these objects of cultural meaning. In writing, I do too. "Waste" is a description that conceals the material relations of capital and property's clutches, inoculating us to how property becomes a grammar of seeing. As a result, pictures, too, misdirect attention from how the visual economy of waste is deeply embedded in perceiving race, gender, and abjection. Too quickly, Blackened environments and Black *people* become things axiomatically in need of "improvement."[12] While it is also true that for these same reasons, picturing these objects could be seen as a form of care (and caring for life), this chapter seeks to foreground the genres of telling (not the visual economies) that emerge in epistemic tension with being made "a waste."[13] In the struggle to not lapse into such traps, I hope to contribute to proliferating alternative ethnographic modes that refuse the twined pornotropic desire to *see* poverty from a comfortable distance and the problematic anthropological pull to *prove* that the poor are "human" too.[14]

FUGITIVE GENRES, FUGITIVE GENDERS

"This ain't no ladies work," Sal said the first time I expressed interest in joining his crew. "The work is dirty and dangerous, no place for an educated sista like you!" Familiar with the contradictory demands of respectability—I was, after all, raised by a Black working-class immigrant mother who taught me to aspire—I could feel how class was operative in Sal's hesitation and probably would continue to be so. Therefore, I didn't push it. But sharing a cigarette on a summer afternoon, the narrative that Sal would speak around me—of cleanliness, of middle-class womanhood—would crack under the weight of Black womanhood's contradiction. "You're supposed to know better than me. Smoking's no good for you," Sal said as he lit the cigarette he had just taken from me.

"I had no idea!" I said wryly.

He laughed as he exhaled. "You're too smart, too pretty. You're going to end up an ugly ol' fool like me."

Sal and I shared cigarettes every time I stopped by the shop. I had become a regular of sorts after I first stumbled upon his archive spilling out onto the street in 2013. A broken but still playable Fender Rhodes sat on the sidewalk with a sign that said, "Play me." I tickled the keys hesitantly and Sal walked over, gesturing to the sign. "Well? Go on, play!" After a few shy bars of "The Nearness of You," I pulled out my pack of cigarettes and Sal, asking to have one, said, "You mind"? We sat on the curb and he told me about Marty, one of the men in his crew. "He can play like you would not believe! Come back, you should meet him. He'll play you anything you want." So I came back. And I kept coming.

But on the day that Sal finally changed his mind, as we sat on our usual curb, a white woman across the street was fussing with her garbage cans. "You know you're not allowed to smoke here! It's the law. I'd really appreciate it if you didn't jeopardize the health of my community." I laughed as we stood up to finish our smoke around the corner, and she muttered something under her breath.

"Excuse me?" I said defiantly. I was exasperated and knew I should hold my tongue, but the way biopolitics paternalistically polices "health" and whiteness (as property) misnames working-class Black pleasure as hazardous to the environment had gotten me in such a twist. I began to lose my temper. "Is there something more you'd like to add?"

The woman looked taken aback, as if she did this often but no one ever spoke back. She responded indignantly and immediately: "What kind of idiot still smokes in this day and age. It's such a disgusting habit. Truly awful." I looked back at Sal, raising my eyebrows and stripping my teeth. "See how it sounds? Hmm?"

As we walked around the corner, I said to Sal, "You know, the thing is, smoking, not smoking, Black people are always conceived of as a 'threat' to the environment. Somehow her recycling her four-dollar Perrier bottle, which I bet you will end up in a landfill, is more righteous than the threat we pose to her 'safety.' It's fucking bullshit! I'm tired of this white environmentalist crap. Seems to me, it's just another way to criminalize Black people and poor people's pleasure."

Wide-eyed, Sal paused, then nodded with a wide-toothed grin. "Well shit, you *ain't* no lady, huh!" Perhaps it was the smoking, perhaps it was the cussing, perhaps upon witnessing the confrontation of a white woman's essential vulnerability to the contagion of Black bodies, what constitutes "womanhood" was thrown into fundamental flux.

I smiled, "No? What am I, Sal?"

Whatever it was, on that day I tipped over the edge of something, perhaps of the whiteness of property revealing my gender or Blackness or both anew, or perhaps it was over the edge of the property lines that make cisgender identities a property of whiteness. Either way, Sal responded, "You're something else altogether, honey! *Something else altogether.*"

Gender is (among many things) a genre though which fabulative tales of the city are told.[15] At the same time that gender (and Black women's genders in particular) is a site of punishment and loss—loss of place, time, community, kin, home—the elaborative practices that spring forth within punishing loss require fugitive knowledge about the unruliness of matter and the contradictions we bestow upon it. Instead of seeking to "settle" matter, as I would learn scavenging with Sal's crew, objects read athwart the sign of waste begin to refract Black genders as a site of matter's possibility, a genre of animation that is often de-animated by property regimes.[16] What if, alongside the way that gender comes to instantiate matrices of domination, it is also a kind of narration? Stories that on the one hand are stories of dispossession, anti-Black violence, and gentrification are, on the other hand, stories about Black women, their labor, and *their meaning* as a genre of Black dispossessed life. Black genders are sites of possibility, forms of elaboration that not only mark "constraint" but force the relentless possibilities wielded in the face of violence to the fore.

. . . .

"Just because we don't mean nothin' to these fucking developers doesn't mean our lives don't have meaning," Marvin said, shaking his head. Marvin was one of the houseless men who worked with Sal to scavenge buildings caught within capital's clutches. We sat in what some scholars and perhaps some viewers might call "debris."[17] But to Sal, Marvin, and Terran, this was evidence of something else. "We have full lives," Marvin said, shaking his head. "This here, is a full life." We were sitting in a building about to be demolished. It was the third Sal heard about that week. Sal had lived many lives in Bed-Stuy. He'd been a fireman, a gut demolition worker, and a scavenger. While these jobs correspond to different moments in Bed-Stuy's history, all of them had taught Sal about the constant destruction of Black life. After retiring as a firefighter in 1995 and as a demolition man a few years after that, he opened a shop in 1999. His "junk" shop.

Marvin lamented as he picked up a pair of socks: "And this here," he said as he fingered the socks gingerly, "is a trace of love." Delicately, affectionately even, he turned one sock inside out to show where it was darned. "My mom used to do this for me. I was always running and jumping over this or that, getting my clothes caught on fences or roughhousing with other kids. I remember this one time, I was at a block party, and my cousins and I had had some of that good punch." Laughing, Marvin continued, "Oh lord, my cousins and I were toe up, play fighting in the way boys do, and I was running because my mama could tell we'd gotten into the punch. I fell flat on my face running away from her and ripped my sock on the stairs or something. She was piiiiissed. You know that Black mama anger. She had just bought me the damn socks too." Marvin sucked his teeth. "I was a stupid, stupid kid. My mom, shit, she might've slapped me around a bit, but you best believe that night she went home and fixed my sock just like these." Showing us the pink thread delicately encircling a tear, he said, "This man was loved by his mama or his girl. He was probably just as stupid as I was too."

Marvin was not the only person to conjure his mother in the dark. In fact, it happened often. On another night, while gathering up some scrap metal, Terran told us a story while holding the side paneling from a refrigerator:

> This fridge look just like the one my mom use to have. Mm, she used to cook the best stew chicken. I'd smell her browning the sugar from my bedroom and run to the kitchen holding onto the doorjamb. I loved to just stand there with my eyes closed taking deep breaths. It's not that the browning sugar smelled good exactly, more like a sense of antici-pation. I would take deep breaths and imagine what dinner was going to taste like. I should've learned how to cook, man. You know, my mom always tried to get me to help out in the kitchen but all I wanted to do was play with the knives. I was so impatient. Well, I sure am patient now. Life's knocked me around some; she was trying to prepare me for it. I should've learned to cook. I should've listened to my mama.

It's easy to understand these stories as some sort of oedipal manifestation, or stories about a loss of innocence or the way that mothers are, even in narrative, sites of social reproduction. It's true, too, that both Marvin's and Terran's stories of their mothers are heteronormative—stories that reen-act the roles of feminized labor as that of caring for or remaking the do-mestic, and masculinized labor as something outside the home. Women darn socks and cook dinner, while the men and boys are somewhere else. It is also true that these stories are fantasies of the domestic, a domestic that is stable and private. Gender, here, is a genre (bios-mythos) through which fabulative tales of the city are told. The fungibility of Blackness casts a shadow of doubt on the genders to which the Black body belong.[18] Gender is a *genre* for the body who, in "flesh," is "that zero degree of social concep-tualization that does not escape concealment under the brush of discourse or the reflexes of iconography."[19] Gender as (a Black) genre does more than constrain bodies through the relations of domination, though to be clear, it does that too. Gender, as C. Riley Snorton and Marquis Bey theorize, is also a cite of elaboration, including elaborating matter.[20]

As these men tell stories about their mothers, another story emerges: the vexed stories of what it's like to be the *child* of "a problem." The enshrining of slavery as an inheritance of the mother ensured that children born to Black women would follow in their legal status of property. These Black men tell stories not only against the violence of *partus sequitur ventrem* and its afterlife but against the violence of turning their mothers into property. By telling these stories of Black women exceeding property, they refract stories of themselves (of Black manhood), too, as more than inheritors of objecthood or more than the violence of being forced to claim hegemonic (white/cis) "manhood." The intimate stuff of pathologized Black life is shot through with the violence of white supremacy; in addition, these stories are intimate knowledge of the way we teach each other to survive the mattering of Black life. As people with mothers who are perpetually held captive, Black men know, too, that gender is a site of punishment. These stories are stories of connection in a world where Black social life, family, and home are constantly being ripped apart.[21] They are stories about the labor of care and the loss of it, about the way that waste reminds people of those that have been rendered a surplus to land (see chapter 2). They are stories about kin, which, too, have been disappeared, and they are stories, perhaps most importantly, about themselves on the verge of disappearance.

Like Hosbey and Roane say of the fugitive ecological knowledge required to flee the plantation, the fugitive labor of poor dispossessed Black men is material knowledge of the Blackening of landscapes. Laboring the waste produced by the spatial violences that discard their homes, their mothers, and their lives turns the environments of dispossession into a different kind of abundance.[22] These men not only know the terrain of destruction—what a rezoning decision means, when it will happen and where—but use their knowledge to forge intimate relations with matter—the unruly, chaotic, and deviant matter of Black life.

The intimate histories spoken in the spaces of scavenging are not always "true." Marty told me privately that Marvin hates his mother because she was never around and blames her for his addiction. The point then is not whether the stories are accurate representations of biographical "facts"; instead, it is that telling stories is a mode of making and theorizing amid dispossession. Within the political economies that encircle and shape *seeing value*, these *gendered* tales of a place being rewritten are also a mode of wrestling with the consignment of Black personhood to the ungendered object—a way to make living room through and across the property regimes that organize nonwhite life for extinction and removal. Against the way whiteness turns property into a grammar of seeing—where value production lies in the expropriation of Black labor, in Blackness's disposability, or as the "outside" to value[23]—these stories are a different way to read with/in the rubble of a continuously dispossessed Black life.

From trash to treasure, then, is a vexed frame. Finding things that have a resale value in the competing cultural lives of a gentrifying neighborhood doesn't mean that the darned socks make it back to the shop. After all, scavenging requires speculating *like* capital: seeing future markets and betting on what potential clients want to hear, itself a speculation about what Bed-Stuy is *becoming*.

> SAL: I think part of Bed-Stuy will always be hood. It's hood in its heart.
>
> MARTY: We make *legends* here, girl! Ooo!
>
> SAL: Not anymore, Squee. We been making a killing at the shop 'cuz everyone is disappearing. You know, when I first started this place, sure, it was about hustling, about making some money, but it was really about helping people out. It was about knowing that people need things fixed and instead of paying an arm and a leg they could come here for some advice and some parts. But now . . . Few years, ten maybe, we not gunna recognize this place at all. It'll be our foundation, our bones, but you won't see us no more. Our history will be buried. It'll be like we were never here.
>
> MARTY: But hood in its heart!
>
> SAL: That's right, the hood'll be somewhere here. Even if they can't see it, we'll still feel it.

That Bed-Stuy will be white confirms the present value of certain things while discarding the meaning of others. This seeing, bound up in what Katherine McKittrick and Camilla Hawthorne call the cartographies of struggle, demands that one "flips" things from Black trash to white treasure.[24] Yet as Sal notes, the stories, the heart, the feel of Bed-Stuy can't actually be stolen. As stories of disappeared Black women attest, conjuring mothers in the rubble of spatial violence elaborates a genre of seeing matter differently. Storying against Black life as waste, these tales exceed how racial capitalism makes matter *matter*—as property and as capital.

Since the 1990s, Bed-Stuy has been a site of intensive criminalization and policing. Famously terrorized by Rudolph Giuliani's "broken windows" policing strategy, Bed-Stuy has been characterized as a place where people disrespect property rights and a place made dirty by Black people's propensity for improper living. When Sal first acquired his spot, the public discussion of city safety relied heavily on the language of cleaning: "A cleaner city is a safer city," Mayor Giuliani said at a news conference in 1994: "That's something that everyone instinctually understands. And something we have to make a big part of our efforts to improve the quality of life in our city. In a big city like this all of us have to learn how to respect the rights and property of others."[25]

Assembling a graffiti task force, a quality-of-life task force, and expanding the budget of the NYPD, the theory of broken windows became the primary logic through which the city recast its own histories of disinvestment and infrastructural neglect as a problem produced by the poor.[26] Giuliani's obsession with cleanup—a project aimed at increasing the city's property values and "quality of life" for the white elite—became, as Neil Smith argued, "a vendetta against the most oppressed—workers and 'welfare mothers,' immigrants and gays, people of color and homeless people."[27] Giuliani's assault on the poor was successful: After he changed the "climate" that "tolerated" how poor people's material practices attempt to remake the city (things like graffiti, panhandling, hip-hop, etc.), in 2001 Mayor Mike Bloomberg picked up where Giuliani left off. In the early 2000s, Bloomberg's administration encouraged a series of rezoning projects and Business Interest Districts (BIDs) targeting neighborhoods of color for "improvement."[28]

But urban improvement not only has a long racist history, it is at its base a colonial project of heteropatriarchal environmental control. In practice, "improvement" is the violent clearing of people whose stewardship is deemed "wasteful" and genders deemed a problem for white property (and its) cis-sexist value(s). As property disenchants matter (and Black flesh), endowing it only with the capacity to be productive for capital, it also imposes a gendered ideology that sorts productive capacity along the racialized lines of respectability. By equating waste with wageless people, ideologies of improvement attempt to erase oppressed people's agency to determine what the imposition of property and gender *means*.

Though heteropatriarchal white supremacy seeks to rewrite the environment, disappearing other ways to read the city, Sal and his crew remake the hood's environmental archive fecund with a gendered material history. In the darkness of places being violently dismembered—life evicted, power cut, and buildings rendered "waste"—storytelling makes matter a site for people to re-member.

Abandoned apartments feel different just before dawn. Haunted, maybe. There's a particular kind of fullness in the dark that swells in the ears, waiting for your eyes to find where the light comes in. And then just before it feels as if the absence of sound might overtake you, it dissipates, turning into a new horizon from which to look, feel, hear, sense; a different horizon from which to move that perhaps offers a different horizon from which *to be moved*—to feel *and* be felt. In the dark, navigating, too, someone's forceful removal and, simultaneously, someone's transient presence, senses are heightened, vigilant even, to the traces of life, violent erasure, and *potential* violence that the wrong move could bring. This embodied mode of analysis doesn't seek distance from waste but moves differently within it. It is a mode that is constantly recalibrated by the signs of life and signs of violence that appear at the edges of classificatory schemas. The disappeared haunt the edges of meaning, making the schematics of violence an informant but not necessarily a claim to truth about how one knows, names, and reads in the dark.

Inside a studio apartment with the power already turned off, Sal sat on the floor holding a picture album: "Damn," Sal said. "How do you 'value' a life? I mean really? What do you do when all that's left are pictures? I never really know what to make of these. It feels wrong to leave them here—to abandon them. But it feels wrong to take them too."

On that first trip to a building on Malcolm X Boulevard, there were pictures on the mantel but so many more on the floor. Cracked picture frames, shattered glass, and tattered images were strewn across the floor. "You believe this? So violent, so fast," Terran whispers to me in the dark. Like forensic scientists, we scanned the scattered belongings and the space between pictures on the mantel. Marvin began to move toward the pictures, unfastening the back of the frames to take the pictures out. He was collecting the frames. Terran began collecting little knickknacks: "These are the things that people can't really take with them, unless they're really meaningful."

I joined Sal on the floor, where he was still cradling the picture album. "This was someone's home," he said. "You always have to remember that."

"That's right." "Sure is." The others respond as they continue to find and quickly assess things. "This was someone's home," Sal says again, this time opening the album.

I was pulled by the images, the loud swell of darkness and then its dissipation recurring in my ears. I strained to find the light while Sal turned each page slowly, the peal of plastic pages cutting through the darkness. At first I thought that Sal could see through the dark; the pace at which he turned the pages *sounded* like he was looking at the photographs, but in fact he was flipping the pages with his eyes closed, running his fingers over each page as if reading the images like braille. It was not the contents of the images he was "listening to," as Tina Campt might encourage, but their presence through time, their imagined origins, the sounds they made in someone else's hands.²⁹ A picture album as evidence of home can be felt without knowing who is inside. Straining against the archive, so to speak, Sal, still with his eyes closed, began to tell a story:

> Did you ever know Sammy on the block? He was a knobby kneed, mean, *mean* old man. He used to sit on a folding chair on the corner with a cane, holding a picture album. Back when Bed-Stuy was still the hood, Sammy would sit on the corner yelling at these young Black men, telling 'em to pull their pants up and don't bring drugs into the neighborhood and all that. He even used to try to hook his cane in the belt loops and pull people's pants up. Anyway, I was a young man at the time, twenty or so. Sammy liked that I worked for the city, felt like I could take care of the block, in case the fires in the Bronx came to Bed-Stuy too. I used to come by the bodega across the street from where he sat, and I'd pick up a juice or a coffee or something for work. And I'd always get one for Sammy too. But one day, he just wasn't there. At first, I didn't think nothing of it until the day after he still wasn't there. And the next day and the next day. I found out that his wife passed, maybe a few weeks before that last time I saw him. And then . . . he just died. Broken-hearted maybe, like he couldn't stand to live without her? There was a little funeral, and even though everyone hated him, everyone came, you know, we're family. We were all sitting in silence at the funeral when all of a sudden, I hear this plastic crinkling in the back of the room. It was one of his sons maybe (some family member) who didn't come around the neighborhood much. Anyway, he was in the back of the room holding the same picture album that Sammy always held. He was turning

the pages with his eyes closed, like he was talking to Sammy. I'll never forget that sound, or the look on his face. It was like, that's what history sounds like. A picture album with no one left to turn it. Someone put this together to hold and turn the pages. This is what it sounds like, history. Sometimes it's loud, but mostly, it's just quiet.

When Sal asks if I know Sammy, it's not a rhetorical question; it's a narrative device: an invitation to a past that *could've* been mine. A history that is shaped by loss as much as it is shaped by modes of reflection. The picture album in his hand was not so much a portal to history as it was the terrain of a speculative imagining, one informed by the rigor of knowing, living, and experiencing Black life. This is what history sounds like: quiet.

Much theoretical attention is given to the ways in which Black life is loud, expressive (and often made public); I also want to draw attention to Sal's formation that what history sounds like is quiet. "Quiet is antithetical to the way we think about culture and by extension, black people."[30] In *The Sovereignty of Quiet*, Kevin Quashie argues that Black resistance overdetermines the analytics of study and analysis we apply to Black life. The "double" of double consciousness has also created a Black subject (on the page) who can only be animated by the struggle against whiteness, a foreclosure that tends to obscure the multiple modalities through which Black living happens. On the one hand, the story about Sammy is a story about respectability, about the emergence of hip-hop in New York City and its aesthetics and the way Black communities fought about Black humanity *through objects*. Oversized clothes = criminality; criminality = Black. This narrative aligns itself not only with other people's stories of the neighborhood, or Black liberals who saw Black youth culture as deviant, but with Giuliani's broken-windows policing. These public sanctions are also the material terrain on which police violence is acted out: it *sees* degradation and acts accordingly through incarceration. Sammy's struggle, perhaps, to avoid the overdetermining gaze of Giuliani's police force was a battle he would never win. The fight about "criminal" aesthetics (be it clothing or the environment) was always about the potentiality of real estate value; eminent domain as a social technique (not just a legal practice of theft) and the settler's ongoing desire for more and more settlement is not a fight won by pulling up Black men's pants.

On the other hand, the story about Sammy is also a story continuous with the history of what Orlando Patterson called natal alienation: the loss of kin and the networks that make "kin" meaningful, the loss of home and the ability to define it, the loss of histories and geographies full of cultural and material referents, and the loss of names that could animate the dispossessed and disappeared.[31] While Patterson coined the term to explain how "chattel slavery's reliance on biological lineage required . . . violent delegitimation of [slaves'] past, present, and future kinship networks," the afterlives of slavery continue to tether Black life to social and material death.[32] For Sammy, the loss of his wife was the cumulative violence of racialized and gendered dispossession, a violence so unbearable he died. Even within the heteropatriarchal capture of this dynamic, Sammy's wife (unnamed) does the work of social reproduction, is the maternal figure who cares for husband and home; she is so central to any precarious relationship to belonging that upon her death, what follows is the death of more than any one person could name. As Sal would say at a different moment, "When Black women die, the whole community dies with them."

As we moved through the quiet, Black women were everywhere. They haunted us, they haunted me. I continuously returned to the question: What does it mean to consider this *not ladies' work*? Ladies aren't supposed to animate the disappeared? Ladies aren't supposed to be the keepers of dispossessed and disappeared history? Ladies aren't supposed to be dirty? Lady is *only* a genre, not a material relation? It occurred to me many years later that perhaps the expansive ways that Black people make and tell history are also a queer embrace of presumptively "degraded" material relations. Perhaps, when your kin networks are constantly severed, and urban policy places impossible demands on you to *be respectable* and respect property, even though it is organized to wipe you away, telling stories of disappeared Black women is a mode of embracing the queerness of Black life. As Cathy Cohen argues, "who and what is queer" in the domain of Black life needs to be expansive, "with an eye toward recognizing and transforming how people live and the desperate conditions they too often face."[33] If queering Black studies, as Cohen suggests, means, "making visible all those . . . who [are on] the margins of society and excluded from the middle-class march toward respectability," then I encourage us to see the wageless embrace of disappeared women as a way of queering Black environmental histories. In other words, the cultivation of affinity, sociality, and kin amid the matters of Black dispossession is a loving embrace, a queer mode of making room for unrepresentable *and* unrespectable Black life.[34]

LOST AND FOUND

As with Sammy's story, stories of loss can also be stories about respectability, heteropatriarchy, and the fantasy of safety if one just takes on the codes of white property as a truth. If, as I established in chapter 3, property and respectability are deeply intertwined, the reverence of property is also a reverence for the racist bios-mythos of the reigning genre of "man" and "woman."[35] The racist biocentrism that inheres in the biopolitics of settlement turns the rights of property into a form of racial and sexual hygiene.[36] Try as "respectable" people might, the adherence to the violence of gender austerity can't intervene in white supremacy's rendering of Blackness as "excess" to property or fundamentally nonconformable. To put it differently, a reverence for property does not benefit the descendants of slavery who are always presumed to be doing gender or relating to the material world wrong. Thus, as chapter 3 would suggest, there's ecopolitical purchase in cultivating an irreverence for property.[37] This irreverence is criminal and/or dirty; it is also a way to displace property's pedagogy and a mode of queering use-values.[38]

In the bathroom of an apartment on Malcolm X Boulevard, Squee knelt before a cabinet holding a pleather pouch. In a kind of squat position, he held his flashlight in his mouth, using the fingers of one hand to pick through the beads he had found. I was standing against the doorjamb watching him touch the beads delicately, picking them up to examine them one by one. "Oh, I like this one, what you think?" The bead was wooden, painted purple with stars around the widest part of the oval.

"It's fabulous, Marty." He shot me a gap-toothed smile. He had popped his dentures out because miraculously, this apartment still had running water and he was going to brush them, and maybe later, come back and take a shower. "You think we should take the beads with us back to the shop?" I asked. He furrowed his brow and shook his head.

"No, no, no. No one will take these, not with the kind of clients coming through. I was thinking about it for me! You think it would look nice on me?" Marvin, who was in the kitchen, yelled something to Marty that I couldn't hear. Over time I would come to see that Marty and Marvin made fun of each other a lot, and though as far as I could tell it was never malicious, the jokes were always the same: According to Marvin, Marty had "girly" taste, and according to Marty, Marvin had "bad" taste in music. Whatever Marvin was yelling from the kitchen, Marty rolled his eyes and responded only to me: "He's just jealous of how handsome I am. He just can't pull off purple!

"Marvin ain't got no style. No taste in music. But for me, it's like the purples, the pinks, the yellows, I hear them just like I hear the blues." And Marty did. At Sal's shop, he would sit outside playing the 1976 Fender Rhodes, and it never mattered if it was "in tune." He'd fill the block with the blues—and sometimes in all blue. People would slow or open their windows to listen. And he loved the attention. Once on a scavenging trip in South Bed-Stuy I found a fedora, then on another trip a hat pin and a feather. And the fedora was purple. After I gave it to Marty, he would leave it at the shop and ask after it when he played because he wanted to be, in his words, "in tune with his groove." I was standing against the doorjamb of the bathroom while Marty continued to examine his purple bead. He stood up, but at six five he had to bend down to whisper something in my ear: "He's not tuned into the same frequency you and I are." I laughed, and he continued. "He's a good man, he's just, you know, a little boring." And with a wink he added, "Solid good is sometimes a little boring."

When I started my first full-time academic position, I ran into Marty on the street. It had been at least three years since I had last seen him, if not more. He grabbed my hand as I was walking by a set of benches across Lexington Avenue and his big callused hands immediately called me back in time—or rather, out of the disciplined time to which I was supposed to be adhering. As we embraced, I realized that he was the only person I'd found again. Learning from those whose lives are constantly subject to the economic depreciation of life meant that I had to learn how to write without follow-up conversations and without returning to people, and how to write alongside and through loss. I had to learn how to conjure, as these men once did in spaces with meanings wiped off the map. Offering him a cigarette, we slid back into a familial relation and the queerness of a waiting embrace:

MARTY: Baby girl, it's so good to see you! I was just talking about you!

AUTHOR: Oh yeah? What were you saying about me, Marty?

MARTY: Well, I just got out of prison, trying to get my shit in order you know. Just got fitted for some new dentures, you know, so I'm feeling good, and I go to find Sal and the shop is just gone! I mean, we knew it was coming. Sal knew it was coming. But *you* knew it was coming; you remember what you said to me?

I shook my head: No. I hadn't realized that anything I said had been worth remembering.

> MARTY: You said, "Marty, it's coming. You gotta find a safe place to land. Developers bring cops, and cops bring developers." And you was right. I got thrown in for scavenging metal and then not being able to pay a fine. I was in for two years. Right before I got out, I was telling someone inside that there was this girl who used to come and talk to us while we scavenged and how you used to say the smartest things. And then you went down South, right? Did you find those landfills and things you was looking for? I'm so proud of you for getting out. You did it. You really did it and became a professor. You let them know when you write those essays that I said you would, that I knew way back when you was just foolin' around in the dirt with us fools. You tell them that I could see your star shining in the dark, like when we was in that one place and we thought we had found some diamond jewelry but it was just cubic zirconium. You remember what I said to you? I picked it up and I said, "Baby girl, this is you. Except you're the opposite, people think you cubic zirconium, but really, you a diamond!"

From what I remember of the trip Marty's referring to, it was neither cubic zirconium nor a diamond—it was plastic. I remember saying to Marty, "Diamonds aren't really my thing." I remember saying, "I'd rather be plastic because I think I'm already *part* plastic," a joke I often made that slighted the racist and cis-sexist assumption that purity is an axiomatic social or environmental good. Also for me, being "part plastic" was a way of saying "queer"—a quiet gesture to the kinds of aesthetics, the forms of monstrosities born of the "pure" categories that demonize Black life, a gesture to the kinds of degraded femme aesthetics often deemed trashy.

But for Marty, "you a diamond" was a way to say something else; perhaps it was a way to refract what I meant to him. Perhaps it was a way to see poetry "as a property of things."[39] A way to note, to *name*, to re-member the treasures of his *Black life*. A way to express care, to animate a Black woman lost and found in the genres of gender, to fabulate a place to which he can't really return but to conjure a queer Black time (or a future) that was (and is) ours. As Kara Keeling's engagement with surrealist theorist Franklin Rosemont suggests, "calling attention to living beings that have been relegated to the status of things" has something to offer how we think about environments.

Wageless and houseless Black people cast as "excess," "unecologic," "criminal," and "hazard" are part of those thingified theorists who "have fashioned movements and cultural forms, including philosophies, out of their sustained engagement with other matter over time."[40]

As Marty brings things close—purple beads, music, his mother, me—he refracts the fabulative genre of gender and the queer potential in making use of objects that property teaches us to believe have fallen into "disuse." Loss, dispossession, the destruction of homes, and the stealing of kin are all processes that create distance. And making risk and toxicity proximate to, and a product of, these same processes makes dispossessed lives deeply entangled with the very material risks of living under racial capitalism. Property (and it's heteropatriarchal wasting relations) offers a seductive promise of safety from risk, the false promise of the settler-colonial project's invitation. But settler-colonial racial capitalism requires environmental theorists to revere property more than life itself, and is thus predicated on the disposability of collective futures. This "use" of environments requires that we don't see Marty—or anyone in Sal's crew for that matter—as forwarding an environmental vision. Not least because it is *not* predicated on a relationship to native lands as ownership, extraction, cleaning, clearing, and dominion. Instead, in the spaces of scavenging, a place where history needs to be elaborated with personal stories to speculate on who might've been without definitively saying what *was*, is a way to make intimacy, not distance; make alternative use, not dominion; make gender a site of possibility, not determinacy, in the environmental devastation of colonial racial capitalism.

The litter of Black life leaves endless possibilities open to interpretation. The structure of condemnation, of toxic capture and becoming fill, turns so much of Black life into the waste on the floor. Though necessarily speculative in their accounts, this crew notices history: retelling who and what counts. Turning litter into a sign of Black life is a way of seeing history in unlikely places: A picture album congeals histories of police violence and gentrification; a darned sock holds affective attachments to dispossessed mothers. Storying these objects is a way of making Black sociality amid a social fabric that is cut over and over again but continuously rewoven by the attention of the dispossessed. In the dark, Sal tends to objects. Attentive to the violence and pleasure of Black living, this tending makes matter *matter*, makes Black *life* matter.[41] And perhaps this tending in its various Black forms—storytelling, speculating, fabulating, and critically straining—is itself a Black ecological practice.

JUNK

There were two kinds of market explosions in Southeastern Virginia in the 1990s: One of them was trash, the other was hip-hop. Home to Missy Elliott (Portsmouth), producer Timbaland (Norfolk), hip-hop duo Clipse, and Pharrell Williams and the Neptunes (Virginia Beach), Southeastern Virginia was also a place for the Black sounds of cheap land—the crunk of Dirty South hip-hop. As deindustrialization and privatization further cleaved socioeconomic and racial segregation in the Tidewater—the Tidewater, too, as something economically and aesthetically distinct from the Hampton Roads—a sonic story emerged from the archives of breathless Blackened environments.

From hip-hop to visual and material culture, Dirty South is a genre of being, a Black mode of telling, perhaps even a series of oppositional gazes with origins as diffuse as the plantation's reaches.[1] Its sonic and visual mandates tell an ecological (and, for some, anti-ecological) story in its blending of the profane and the land, dwelling and apocalypse, grit and spirituality. This story emerges within the "cacophonies" of dispossession and rewrites waste as dispossession's attendant condition.[2] If, as Pavithra Vasudevan argues, "Plantation slavery was a laboratory for studying and managing 'unruly natures,'" then plantation slavery also produced a future of "unruly" matters materialized in the atmospheres of racism and the knowledges that lie therein.[3]

In *Development Arrested*, Clyde Woods argues that the blues is an epistemological critique of development (i.e., plantation relations) that insists that we reckon with the representational distortions that were "forged in the bowels of the plantation."[4] Among them he names the way "the reification of urban street culture *detached* explanation from its blues-folk roots; the analysis of African American social life as *urban* became definitive and the rural south became derivative."[5] No more iconic moment of hip-hop in the 1990s better showcases this Black North–South divide than Outkast's often-cited 1995 speech at the Source Awards in New York City, where, while being booed off the stage for winning Best New Rap Group, Outkast yelled into the bowels of racial capitalism's capture of Blackness: "The South got something to say."

Blackness as "urban" had also become something Northern, a discourse that was made material by the waste industry's synonymous capture of the post-plantation lands of the South, stained by, if not stuck in, history. In addition, the deindustrialization of Black urban centers in the North gutted social services, leaving Black people wageless and contained by deteriorating public housing. The images of urban decay in the postindustrial decline of cities like Detroit, Chicago, and the Bronx became evidence that the expansion of incarceration was necessary. But as Rashad Shabazz points out, subject to the carceral forces of containment—spatial isolation and incarceration— poor Black and Brown youth "remixed the geography of containment and poverty into one of freedom and 'furious styles.'"[6] Hip-hop turned the material realities of the Bronx into "fertile ground for the development of art and culture."[7] This is an important injunction to the criminalization of Black material cultures that see hip-hop, graffiti, baggy pants, and trashy clothes as a wasteful form of deviant cultural production. The struggle to survive the spatial and ecological violences that burden Black people with toxins and symbolism often requires remixing matter into an ecopolitics that cannot afford to distance waste from freedom.

By the end of the 1990s, there would be twelve private mega landfills in Virginia, eight of them in Southeastern Virginia alone. Waste contracts in the North supporting the excessive presence of landfills in the South produced a geography of waste continuous with the region's environmental history of slavery.[8] In many ways, "the dirty history" of the South was literalized as a racial geography of waste that implied that Black people, particularly in the South, were behind or outside time. Indeed, representations of Blackness were forged in the bowels of the plantation, and *Black* and *trash* became redundant on those same lands. The waste industry itself, and landfills in particular, are colonial strategies for mitigating environmental risk, burying the risks of settler-colonial racial capitalism in the flesh of devalued bodies, nonhumans, and land.

If managing waste is in no uncertain terms the management of the geographies and flows of racial capitalism, then securing waste contracts relies on the constant sucking of land, ever expanding into new "frontiers," including the ever-possible frontier of the Black body. From the slave to "a proper sink for pollution," the Blackened environments of toxic capture justify and simultaneously disavow the violence of conquest.[9] Waste is in its proper place near Black bodies because Black people, too, are an excess to land. In other words, as Atlanta-based rapper Goodie Mob's prose in the 1995 hit song "Dirty South" suggests: "See life's a bitch, then you figure out / Why you really got dropped in the Dirty South."

So, I return to Dirty South hip-hop as a genre of being to make a point about Black material culture specifically and more broadly. As the waste industry was turning the South into fertile ground for the North's *junk*, Black people subject to waste's toxic capture were also remixing the environmental outcome of capitalism's death-dealing logics. DJ Screw's distinctive "chopping" and "screwing" of time re-composed the "outside of time" that Blackness in the South had come to represent. The slurred rhymes of the drug-induced trap house remixed the slowness recurrently ascribed to Black rurality, inviting pleasure into the ecologies and geographies of abandonment. Dirty South hip-hop is tonally and aesthetically responsive to the distinct forms of material violence that shape and name the South, but it shares an ecological injunction with sonic production in the Bronx: to live, and it is an ecological achievement. In other words, Black material culture is an environmental response to the eco-logics that devalue Black life.

"Black musical space," as improv scholar James Gordon Williams argues, "is rooted in the materiality of Black people's lived experiences."[10] What if we were to take the materiality of Black people's lives seriously, to see the concentration of environmental risk as a condition of Black life, to see Black material cultures as also ecological—even if the ecological politics are ones that marshal unfamiliar stories to those of a utopic "nature" and pristine environment? What if Black music and other forms of Black cultural production (graffiti, aesthetic adornment) speak from and back to the ecologies from which they spring? What if turning "junk" into *crunk* is an ecological experiment, with geographic condemnation and hip-hop an environmentalism born of discarded living? What if when Missy Elliott samples "I Can't Stand the Rain" in her hit single "The Rain (Supa Dupa Fly)" (off the 1997 album *Supa Dupa*), she is talking about the weather (of anti-Blackness) and the way capture (toxic or otherwise) *always comes*?

Conclusion

FICTIONS OF FABULOUS/FABULATIVE ETHNOGRAPHY

> The proposition here, against all liberal universalisms and scientific positivities, is to insist that we do not yet know what a human outside an anti-black world could be, do, or look like. The critical poetics of afro-fabulation are a means of dwelling *in* the shock of that reality without ever becoming fully *of* it.
> — NYONG'O, *AFRO-FABULATIONS*

In June 2015, there was a series of unexplainable fires within four blocks in Bed-Stuy. The curious fires sparked a gossip chain in the neighborhood, connecting kin (fictive and natal) across space and time: *My auntie's sister's cousin Ester who lives in Jersey is married to a guy who is a construction worker for the contractor that's working on that building.* Or, *Well, my boy's girl used to go around with some kids who smoke weed in the lot across from the building that just burned down.* These links to and from, within and without, the neighborhood swelled with speculation: *You know, I saw that same construction company on Gerry's block put up a sign a week before the fire.* Or, *Dusty said there were some shady white men lookin' up to no good on the next block the night before that other fire last week!* The swell of guesses, conjectures, explanations, and ad hoc neighbor-to-neighbor analyses yielded little in the way of a consistent story. But every guess, every analysis, every conjecture, and every explanation was as true as it was false.

The apartment that I lived in from 2013 to 2015 was above the only Black-owned pet store in the neighborhood. Adjacent to the building was a Black barbershop run by two men who, when the storefront was empty, sat outside polishing their brightly colored Harley-Davidsons. A few weeks before our block became the next scene of speculation, I sat on my front steps sharing with them speculations of my own: "That building down there"—referring to a paint store at the end of the block that had been out of business as long as I had been living there—"has been empty for too long . . . I don't know, seems like bad news."

"Ma," K said, polishing his pink motorcycle until he could see his reflection, "it's all bad news. Truth is, none of us is ever going to know what happened for sure, but *we know what really happened. We know what's happening.*"

After each fire was contained, no one went back in. No articles appeared on local news sites. The buildings were vacated rapidly, boarded up, and scaffolded. Just as quickly, signs emerged. First, a New York City Department of Buildings vacate order. Then, a condemned-building notice. Then, a construction notice, followed by a No Trespassing sign. Finally, the last sign, always a projection of a "better" future: a development company name, a rendering of a high-rise that did not yet exist, with amenities that no one *currently* in the neighborhood would have access to (a covered parking lot, a private gym, a rooftop garden). The order of signs is not insignificant. It tells us something about the order of truth—including who and what claims are considered real—and the fictions that govern representations of the not-of-this-earth ghetto.

As K polished his hog, I looked across the street at the boarded-up brick building. It had been empty for a few months too. The old men who scratched their beards as they played cards or sometimes chess were also gone. The gray card table and matching folding chairs had disappeared, and along with them the gallon jugs of lemonade or sweet tea from the Jamaican corner store; the jerk chicken lunch special containers filled with chicken bones, which would accumulate in a pile left for one of the neighborhood bodega cats; the canes that they would rest gingerly on the backs of their chairs; the young boys who would sometimes get a good talking to on their way home from school; and the guy on the bicycle who would change the music to old soul when he slowed by the table. All gone.

A number of revitalization projects—the Bed-Stuy Gateway Business Improvement District, Restoration Plaza, street-scaping projects championed by the mayor's Office of Economic Development—had brought more and more attention to Bed-Stuy's "potential."[1] While betterment projects bring chains to main thoroughfares, a future of economic development brings real estate speculators to "empty" spaces. Lots overgrown with grass, insects, rats, and weeds and adorned with No Trespassing signs are presumed empty and uncultivated, "a waste of space." But *empty* and *uncultivated* are not analogous. Rather, they are colonial descriptions/equations of place that inaugurate settler-colonial racial capitalism's development schemas. While the development that will take place is a matter of historical, material, and sociopolitical specificity, the outcome is structurally generalizable: dispossession of many kinds.

Outside the metrics of private property, *uncultivated* suggests something full of entanglement, density, liveliness, and uncontainable life, not emptiness. But within the colonial eco-logics of racial capitalism, the uncultivated is an empty space into which colonialism's violence imposes "life, liberty and the pursuit of property": a contradictory imposition to which Black and Indigenous livingness is forever yoked. This colonial eco-logic subtends all types of capitalist development, including the projections with which Bed-Stuy's future is now entangled. Apprehended as waste *and* cultivatable space, Bed-Stuy's landscape is an ecological gold mine because, according to the order of signs of truth, Bed-Stuy is not ecological at all.

Still polishing his bright pink Harley, K shook his head. Contemplative, he groped around in the reaches of his mind for something. "What's that game, sis, the one about lying and telling the truth?" he finally says. He was searching for a metaphor. A vexing/complex task for someone who endures being named and misnamed in the pursuit of maps, territory, and private property. Metaphors are often mobilized to kill Black, poor, and colonized people on the page and plan for social death. K, to my mind, was looking for a metaphor to describe the false terms of anti-Black colonial fictions—fictions that condemn Black bodies to their exchangeability and *use*. However, to story that which has always been disavowed is also to reconfigure narrative device altogether. It's to call down the service that Blackness offers to the white literary imagination and disorder the order of signs to speak other truths, which sometimes includes making new signs to tell old stories, recycling old tales to tell something true.[2] K is working in a tradition that recombines old fictions to story the ecocide and genocide that continue to shape the dispossession of capital's migrating futures (slavery's afterlives).[3]

This tradition has many iterations and it appears all over this book. I've traced it in the margins, the songs, the landscapes, the literature, the footnotes, and notably, in the music that forms the backdrop to conversations in this book. This tradition is the forms of protest that art (and theory) has always afforded, allowing the dispossessed to improvise where wasting relations continue to devastate life. We might call it Afrofuturism, we might call it jazz, we might call it the spatial remix of hip-hop, and we might call it a blues epistemology commenting on under- and uneven development.

This practice goes by many names. My intention here is not to rename it, but instead to surface it as an *everyday* practice shared readily with—and by—me.[4]

In the fog of the city's summer heat, I stumbled for the name of the game too. "Telephone?" I queried. Even in retrospect, it feels like an accurate metaphor for the haze of speculative capital around us. Distortions mobilized by capitalism's decrees, the story—never quite citing its sources—is passed, changed, and passed again. But the "speculative" has multiple uses; it is both a practice of Black storied conjecture and the process by which the accumulation of the fungible Black body is turned into venture capital.[5]

"No, no, not that one," K says, "I mean the other one." I know what he means, but I can't remember until he says, "You know, where you have to tell people lies and they guess what's true?" And then it clicks.

"Ooooooh, two truths and a lie!"

"*Exactly!*" K says emphatically. "This," he says, pointing around to the neighborhood where the card tables used to be, to the pharmacy at the end of the block now boarded up, to the apartment that I was about to be kicked out of, "this is just like that. We don't need to know who set the fires *to know who set the fires*. Truths and lies, Ma. Truths and lies." What is "proof" of anti-Blackness in an anti-Black world anyway?

FABULOUS FICTION

Being Black requires being an artist of sorts, an informed speculator. Perpetually in the shadow of humanist disciplines (for which speaking or describing Blackness serves as a postscript) and simultaneously serving as a universal scene or staging ground for spectacular violence, engaging Black lives always means engaging fiction. Sometimes this fiction takes the shape of "the stereotype" (the mammy, the slut, the welfare queen), sometimes it takes the form of environmental thresholds that determine when and how much toxicity affects ecological health, sometimes it is the "un-ecologic" that casts Blackness as unnatural or outside of nature-social relations, and sometimes it is the putative sexological/raciological claims of science itself.[6] The overrepresentation of Man to which truth claims adhere make racism, cis-sexism, and classism tenets of colonial fictions. Thus, as scholars in particular, we are required to engage these terms, no matter how distorting or wasteful. Writing against (in other words, attempting to refuse) an apparatus that *wastes* requires other methods for telling some dark truths. These methods already exist within Blackness's complex capacity to signify otherwise than the tools of reason and representation might have it.[7] Over the course of this book, I have tried to employ speculation, improvisation, and fabulation as a way to wrestle with regimes of truth. I do not mean to suggest that Black people are somehow not real (or the corollary misunderstanding, one that fascinates anthropologists, that our interlocutors only lie), but I do mean to suggest that just like the world-building power of fiction, being Black (in an anti-Black world) means being something *of* fiction.

The four fires that went unreported in Bed-Stuy included a fifth. That final fire of the series (though not in the summer) resulted, unsurprisingly, in Black dispossession: the disappearance of the barbershop, the pet store, K and me. This book, however, is not about my personal story of loss (though it is informed by it). It is also not about the legal case that ensued, the allegations brought forth in court against me as somehow to blame for the loss of a material life I could barely afford to begin with. Though the truly scarce settlement that required my contractual promise to never disclose "the details" prevents me from naming the parties to the case and writing *the truth* of what happened, this book is still not about that truth. Instead, it is about the ecological truisms of Black life, the ecocide of white supremacy, the materiality of whiteness as property, and the fabulative or fabulousness with which Black eco-grammars speak the terrors of toxicity and the possibility of living *anyway*. This book asks a different set of questions of ecology, or rather of the ecological imagination that makes up "the other" and insists on the anticolonial eco-grammars leveraged in the interstices of the scientific/propertied imagination.

Property is a genre of living. And, as I have argued, requires the elsewhere of risk, waste, trash, toxicity, and the toxic capture of Blackness. It is a rejoinder to the Lockean requirements of transforming nature (i.e., native territory) into (white) property. But I want to return to this question of the natural, for alongside nature awaiting "transmogrification" is the "Blackening" of life through the transformation of matter.[8] Transforming earth through settler racial capitalism's architecture—be it draining a swamp to expand the plantation or obscuring the location of a water well in order to expand a landfill (see chapter 2)—has and *will* have long histories. These histories will live on in the unpredictable alteration of earthly entanglements, including indeterminately in people's bodies, tethered to economic processes that are obscured by their scientific naming, diagnosing, treating—and lack thereof. Moreover, these histories will live on not through their visibility but through what they reap in their wake: the continuous expropriation of Indigenous homelands for new and often toxic industries that criminalize and dispossess those same people who become a colonial frontier of degradable flesh. The Black ecological struggle is as much a struggle to *be* as it is a struggle with the *terms of being*.

In this final chapter, I take a cue from the boys at the barbershop to honor the way dispossession turns lies into truths and truths into lies. Across the chapters, people like Betty, Sal, Marvin, Squee, Mary, Jane, and Terran have been central pedagogues of a different environmental imagination, one in which there is no ecology where slavery and Indigenous dispossession haven't already scorched the earth. They are also people for whom dispossession has made them, for some readers, epistemically untrustworthy, perhaps even unlikable. Whether or not you trust them (or me for that matter), what you must understand is that dispossession is a place of discards and discarded living requires improvisation. As such, it is a place where the struggle over "a description of existence" is an ecological practice, and survival is a critique of colonial "narratives of science and history" that misname, alter, and rearrange our more-than-human world.[9] It is a place that is a condition: the condition of always being *excess* (to land) and a reservoir *indexing* property's ecological violence.[10] But it is also the place of the Black poet who peers beyond the horizon of thought, and in the case of this book, those poets have lives that property can't imagine (or can only imagine as criminal).[11] Elsewhere, people *know*, "without [the] modern categories" upon which whiteness's material authority rests.[12] Elsewhere is an abundant store, a fabulous capacity, an ecological requirement to improvise, experiment, and herald something else.

Experimenting with the modes of speculation that people use to survive, I return to some of the critical scenes in this text. Ethnographic writing is staged, narratives are motivated by authors, and theory is done sometimes at the expense of the conversations we've had. And sometimes we, as ethnographers, lie.[13] Particularly those of us who have "partially escaped" but are supposed to ventriloquize the subalterns, we carefully leave things out in attempts to refuse "the damage centered narratives" that anthropology mines and collects.[14] We change people's names, we change dates, we leave out the drinking or the drugs, but most often we leave out the violence that we witnessed or that befell us while doing the work. The experiments that I pose in this chapter don't hang together coherently under a formulaic methodological intervention. They are, as Kamala Visweswaran once said, "experiments, not programmatic solutions for the practice of feminist ethnography."[15] Fundamentally, her landmark essays ask us, "How does ethnography change when informed by different theories of the subject"?[16] If, as I've suggested over the course of this book, the other modes of being emergent within the dispossessed subject house ecological knowledge, how might we attend to that *ethnographically*? What's at stake in the writing, particularly if, as Ryan Cecil Jobson demands, "to let anthropology burn permits us to imagine a future for the discipline unmoored from its classical objects and referents," enables us to "abandon its liberal suppositions," and, as the epigraph of this chapter asserts, allows us to find ways of "dwelling *in* the shock of that reality without ever becoming fully *of* it."[17]

Instead, these experiments are ways of dwelling by way of retelling things, places, and times that we've already been/experienced. Sometimes that requires restaging scenes that we've already encountered to encounter them again. And sometimes that means letting contesting versions of the truth sit incongruously with one another and without "resolution." It is to call upon other modes of writing not simply as "poetic" or "polemics" but as *poethics*, a way of engaging with the "whole range of possibilities for knowing, doing, and existing" in everyday Black life.[18] My work was full of music, joy, laughter, drugs, dancing, and stories all the while I was collecting stories of loss. It was full of documents, toxicity, sickness, and sadness. It was full of life. There is no reconciling these things; there is only figuring out ways through style and genre to make them part of the experience of reading about Black life that is all too often occluded by its representation. People's livingness exceeds the page, so what is the page for? As Audre Lorde says, for some of us, "poetry is not a luxury"; it is a mode of survival.[19]

FICTIVE KIN

It was on my first trip to Suffolk that the first line of Octavia Butler's *Kindred* came back to me. Right before I left for Virginia, I had just finished it. It had not occurred to me why I would reach for such a text before embarking on this lonely trip south, to the far-flung landfills to which New York City's waste is diverted. In truth, I didn't end up in the right "place." Rather, I ended up in a place that demanded that I understand how impossible waste is to trace. It didn't take long to become disillusioned by whatever questions I initially thought I should be attentive to. After all, my "field"—in anthropological parlance, as Savannah Shange says of ethnographic research—was not very far from another set of fields: the plantation.[20]

The first time I stood in a cotton field, it occurred to me why it *hadn't* occurred to me yet. Truthfully (or is it?), I was being haunted. And the more I tried to understand why, the less sense it made. It wasn't until I got back to Brooklyn—the same year that Trump was elected, the same year that Sal's shop finally disappeared, the same year that a series of fires inexplicably displaced an entire block in the span of a few months—that I understood what Dana meant when she said, "I lost my arm on my last trip home. My left arm."[21] I realized that after all Dana had lived through on her time-jumping journey, she *had* to lose an arm to come home.

The girls at the motel asked me over and over again where I was from. I kept answering, "I'm a graduate student in New York City." It was never a good enough answer. I stood out in a way that I could never quite identify, and I wrote about it constantly in the margins of my notebooks. At the grocery store, Black women would squint when I walked up, like they were trying to focus a lens; like I was blurry, a puzzle, otherworldly; like they had to narrow their eyes to see if I existed. I didn't, really. I was malleable. And the more I said about myself, the more malleable I became. *OK, but where you from from?* After she'd asked me a few times, I told Jane I was born in Tobago. "My mom had me nine days before her sixteenth birthday, and we immigrated when I was a little over a year old."

"What?! That's crazy. I bet your mama is a famous model or something? She gotta be like super-duper smart for you to end up all the way in New York City and grad school and all that! Oh, and you know 'cuz she's so beautiful, she caught herself a rich man and everything to help raise you right. The crazy part about all of it is you ended up down here with a bunch a hookers!"

I remember the first time I was hit on at the grocery store in Norfolk. A white woman approached me in the line, and she looked like someone of a different time. In her polite Southern drawl, she asked, "Excuse me, where are you from?" I hesitated. First of all, I had been in Virginia long enough to understand Southern politeness to be a form of civilized surveillance. Second, I was in sweats and I was wearing Timberland work boots. After being followed by cops in Portsmouth after a gas station attendant called the police on me when I missed the No Tims, No Hoodies sign in the right-hand corner of the window, I had also come to understand that the Tims I bought in anticipation of observing landfills (which I was never allowed into) and my sweatpants were a sign of Black deviance, an aesthetic associated with Black criminality. Third of all, the question, "Where are you from," in the context of Trump's growing popularity in the lead-up to his first election had a particular legacy in my life in which disarticulating my-self from African Americanness was disciplined, lonely, and not quite true. After all, even if I still carry my green card anxiously, am I really "from Tobago" if I haven't been back since I was thirteen and my accent comes out only when I hit a certain register of anger? Growing up, when the Black kids would ask me *Where are you from*, it was a way of marking class even though we lived on the same block, in the same apartments, sometimes even with the same Children and Family Services agents at our doors. It was a way of saying *You don't belong here*—you don't act like us, look like us, sound like us. But when white people asked, it was motivated by an assumption of where I *did* belong: somewhere in Africa, or sometimes Brazil. Needless to say, when this white woman approached me in the grocery line, I was alert to the possible meanings—all of them painful, none of them playful.

When I didn't respond immediately, she said, "Oh, darling, I didn't mean to startle you. Truly, I'm just taken by . . . by your beauty. You're stunning, do you know that? Are you down here in the South alone?" Again, somehow my out-of-placeness, or perhaps my "northernness" was of note. I couldn't help but think, *Is Virginia really the South*? The person by the Target checkout counter smirked with raised eyebrows, shaking her head ever so quietly like she knew what was about to happen. "I'd love to show you around sometime . . ."

I did make it into a landfill once, though. Not onto the landfill ex-actly, but I slipped my way between yard dogs on their way into the SPSA regional landfill in Suffolk, Virginia. I had been building up some con-fidence after multiple promised trips had been thwarted by the SPSA re-gional landfill manager. I had noticed a turnoff along Route 58. After being followed for miles by cops from Portsmouth, I instinctively drove to Suffolk—maybe to visit the Underground Railroad site, maybe just to have a meal at the diner in town often frequented by Black truck drivers. The turnoff was quick, and I had to drive in circles a few times before I was brave enough to take the sharp right turn heading West from Portsmouth, but I finally drove in and found myself accidentally en route to a weigh station. Frankly, I almost caused an accident driving over the grass median that separated the weigh station from the workers' parking lot, but I made it without anyone but the garbage truck drivers noticing. I parked the car and was immediately confronted by landfill workers who noticed that something was out of place.

"We don't do drop-off here; you should know that living here." Although by the time the young man saw my face he had trailed off. To him it was obvious I wasn't from here. "Where you from?" I gathered up the speech I had been preparing about doing research and wanting to learn more about how landfills worked. This was, after all, what I was told after multiple cursory phone calls. After a moment, he said, "No, where are you *from* from?" This time I realized that he was noting that I wasn't from Suffolk. I had violated something about space and time, race and gender, class and place, driving up to a landfill.

There was a second Black man now. He had strolled up when he noticed the younger one sticking around beyond the "we don't do drop-off today" conversation. After overhearing my speech—"I'm a student doing a project about waste management, and I'm just really interested in what you guys do and how landfills work"—he looked at me, puzzled. "Why? What kind of student are you? Ain't no students around here doing work like that." The younger one shrugged his shoulders, willing to at least entertain what I was saying. The older man was far more skeptical. "Are you with the CIA"?

So I went to the peanut factory, not Planters anymore but where it all began. Planters Peanuts is where I met George Washington Carver, or his legacy anyway. Not inside it but across the street, watching Black men slowly march to the factory's whistle. Standing at the railroad that bisects Suffolk into its "Black" and "white" parts, I felt time in a way that "others wouldn't think was so sane."[22] I watched as their boots kicked up dust, and I wondered if it was toxic. I wondered what these soils were like before Carver's project had been wrested from him to rearticulate this place not as the place where slavery began but as a place with "no history." I wondered how peanuts became this part of the Tidewater's savior and demise. I wondered if I could ever eat a peanut again.

I came back to Brooklyn, out of time and out of sync. In ways that Dana knew, I also found kin. I ended up with a bunch of hookers who offered a refuge from the plantation's haunting, an invitation into the trashiness of Black femmeness (not just its aesthetics but also its ambiguity)—a place of non-representability, a place to be forged and rewritten, a place to rewrite the meaning of alienation and Black ecology. The violence that makes Black women unthinkable does steal, but trashy Black girls know how to steal back.

So I might have come home missing a limb, so to speak, but I also learned that Black femininity was deliciously unnatural, artificial, trashy, perhaps even otherworldly.

MARTY BY ANOTHER NAME

Marty had style. I mean, real style. He looked as if Jimi Hendrix and Marvin Gaye had a baby and added just a sprinkle of Dapper Dan.[23] I'll never forget walking up to Sal's shop on a spring morning to find Marty tickling the piano keys, wearing bright blue striped bellbottoms, no shirt, and suspenders. I was shook. If I had already met Betty, I would've told him about her and said that Betty would've said he looked fly AF. He could play too. Damn. I still hear him sometimes, like a memory wafting over Brooklyn.

Squee was also born in Virginia, and sometimes, like many in the neighborhood, he called me Slim. Sometimes Squee said he was from the Caribbean, and sometimes he said he was from the South. Other times he said he was born in Brooklyn, and I'd even heard him say, "I was not born from this world at all." Squee was six five, and his hands looked as though they could palm a basketball effortlessly. He told me he played growing up, but some of the guys in the crew told me that wasn't true—Marvin, in particular. "You know, Squee has had it real hard. He bounced around a lot from foster home to foster home, and I don't know all the details of his life, but I know he had it hard. You can just tell sometimes because of things he says. It's like, sometimes when people say contradictory things, you know it's a way of twisting hard truths. You know what I mean?"

Marty told me that he had an alter ego, another person growing inside him: "His name is Marvin." He told me that the Marvin growing inside him was a man with a soul, a man of music, but a man with devilish urges as dark as his blues. "Strange, you know, 'cuz Marvin"—now referring to Marvin in Sal's crew—"he's like my brother from another mother. We bicker sometimes, fight just like it's true. But I don't pay it no mind because I know he's just looking out for me, even if I think he patronizes me sometimes. He doesn't understand alls I got going on inside." In many ways, Marty and Marvin were opposites. Marvin was "a man's man" in his own estimation. And Marty, in Marvin's estimation, was "a bit fruity." To Marty, "Marvin's like solid wood. He's sturdy, strong, and might be weathered but can stand up straight in a storm. But he's also bound to break. One of these days, he's going to break right in half, and it won't be good for nobody."

"You know why we call him Squee, right?" Marvin asked me rhetorically. Marvin and I were walking to meet up with the crew. Marvin had been late, and I said I would stay at the shop until he got there. Marvin didn't like to talk about himself, but he sure loved to talk about Squee:

> He's so tall and skinny, he can squeegee his way into any crevice. It's truly amazing. I once saw Squee climb up the rafters of an old building, in a place where we couldn't bring a ladder, and Sal thought we weren't going to be able to get nothin' out. Squee said, "Hold on, man, let me try something," and this fool starts climbing the rafters! We were all screaming, "Marty, come down, come down, you're gunna break your damn neck," but he just kept climbing, like a kid climbing a tree. Then we all started shouting, "Go on, Squeee!" and that's where the name comes from. Squee can climb anything, and squeegee in anywhere! It's wild!

Marvin told the story with something more than delight, with something more like pride for a little brother or a teammate, a story that became a legend.

The Saturday morning that I walked up to find Marty playing the piano, dapper in his tight blue pants, bare feet, and bare chest, I folded my arms over the back of the piano and placed my chin where my arms crossed. "Anyone ever tell you why they call me Squee?" Marty asked as he played. I shook my head while humming what he was playing, Marvin Gaye and Tammi Terrell's "Ain't Nothing Like the Real Thing."

> 'Cuz we once found a comic book in one of these building about a kid who'd kinda been abandoned by their family. It was a sad story, the kid's mom was addicted to drugs, told him all the time that she hated him. I don't know where his pops was, but all the adults in his life thought he was garbage. Anyway, I was the one who found it and I took it, I didn't tell no one. I shoved it in the back of my pants and covered it with my shirt. But when we was leaving, Marvin saw the title when we was walking out. I never let him read it, but the comic book is called *Squee*.

Terran had a different view of Marty, one that included Marvin. "I think they really, really love each other. It's deep; they're boys through and through. They've both fallen on rough times and they help each other out. Marvin is always finding food and stuff like that and saving some for

Marty. Marty is so kind, you know how he is, wouldn't hurt a fly. He's gentle, and Marvin isn't gentle. He's not, like, violent by any means, but he's also not gentle. He'd certainly fuck someone up if they ever touched Marty, you feel me? And Marty, well, he makes us all *laugh*, and he can *play!* He's always singing and brings the joy, you know; he always puts us all in a good mood. You know that's why we call him Squee, right? You know how in books when someone's really excited and they write the sound people make like *Squeeeeee.*

TESTIFY

At Sal's shop, doors could testify. "Yeah, I remember his mama. She was strict, ooh, she would tell him to cut his shit out, not to sling dope and all that. But he was out here running the neighborhood." Sal was telling a client a story about Biggie Smalls, who grew up just a few blocks away on St. James Place in Bed-Stuy. "So, this one day, I'm coming back from work and I hear his mom on the step yelling, *Christopherrrr,* 'cuz you know, that was his real name." The client nodded, of course, of course.

> So she's yelling *Christopherrrr,* and you know Loretta, that's his mother, was a God-fearing woman. Church on Sundays, prayers, and all of that. So Biggie kept his hustle quiet because his mama would be having none of it. Anyway, I'm just standing on a street corner when he damn near runs into me with God knows what else he had in his pockets. But get this, he had a bundle of cash THIS BIG on him, just fell out of his pocket onto the sidewalk. Lord, it was A LOT of money. But you know, the thing about him is he just kept giving all that money to his mama. Most people don't know, listening to his rap, he was a good kid. He always took care of Loretta, 'cuz that his mama. He was a good, good kid.

Sal pointed to one of the two doors that the client was deciding between. "This one," Sal said, "is part of Bed-Stuy history."

"This one?" the client asked, wide-eyed. "It was Biggie's door?"

Sal nodded, the way someone nods right before they praise the lord, maybe even the way Loretta would nod at church, praying for Christopher.

"If these doors could talk," Marty said. He was standing in the corner shop.

"Can't you feel it?" Sal asked. "There's something spiritual about it."

"Testify," Marty shouted as Sal spun the door into gold.

The client was enthralled, caught up in the Black rapture: "I just love how Black history is everywhere here," the client said. "You can really see it shine through the neighborhood." Marty and Sal nodded like they were in a Black church, like they were praising this white client like *he* was God himself. After a moment of silence, Marty began to ready the door for sale: He polished the knob, made sure the screws were tight, and double checked for paint chips.

As Marty checked the paint, he made a face, disappointed in the color. I had watched Marvin and Marty and Terran strip, sand, touch the grain, and sand the door again. I'd watched them debate over paint colors. Marty wanted something bright, of course; Marvin didn't care. Terran, who was often assertive but calm, chose red. And that was that. While Marty ran to the back of the shop where they kept rope for transport, the client was telling Sal about his plans for the place he'd just bought a few blocks down on Gates Avenue: "Yeah, we're gutting it. I really want to build something that lasts for my family, in a place with real history. If feels really important to me that I don't take the neighborhood apart, but that I honor it. You know, I mean, I'm a white guy, so it's important for me to do that work. I'm so glad to know you guys are just down the street. I'll definitely be back! This door is really something."

I was with them the day they got that door. We were scavenging a condemned building on Quincy Avenue. The place was right across the street from a building that only a few years later would have a Biggie mural on its wall. Not only would the mural loom large in Bed-Stuy, it augured the multiple murals memorializing the Notorious B.I.G. that were to come. But in 2017, after only two years, it would become a point of aesthetic contention. The community art organization, Spread Art NYC, and the two artists who painted the *King of NY* mural tried to save it. They offered to pay $5,000 to the landlord to keep the mural, but the landlord countered with a lease agreement of $1,250 a month to preserve it. In an interview on a local neighborhood blog, the landlord said, "Why should I keep it? I don't even see the point of the discussion."[24] In a surprising victory, however, the artists won the battle of meaning. Crown tilted, lips slightly pursed, and gold chain partially obscured, Biggie's head looms bodiless and imminent, watching the neighborhood.

Eventually, the client came back with questions. Sal recounted the interaction while we shared a smoke. "The guy come back upset," Sal said, rolling his eyes. He took a deep drag, paused, and blew out a cloud of smoke. "Apparently he told a friend that he bought Biggie's door, and his friend didn't believe him. He came back asking for proof. I said to him, 'I am the proof. I've lived here my whole life. I've seen these kids running around the neighborhood, all of them, Lil' Kim, Jay-Z, Chris Rock, I knew their moms, I broke up fights, I was here. If that's not proof, I don't know what is.' This stupid motherfucker, what does he want—me to go dig up Christopher and give him a fucking finger?!"

One mural, two murals, then three. Do or Dine and then Do or Dive all monuments to Black history. A brownstone with Biggie's door. Sal's shop is no more. Fabulation is a response to condemnation, when the lines of the neighborhood are still red. "I'm ready to die," the Notorious B.I.G. once said. "Shit is real and hungry's how I feel, I rob and steal because that money got that whip appeal."[25]

SWAMP THING

"I'm something of a fiction," she said, winking at me while high one late afternoon in her motel room. Betty had successfully managed to collect enough oxy to distort, or maybe bring herself into alignment with, her relationship to pain, toxicity, and time. Growing up, she had always struggled with the use of one of her arms. "It's my dead arm," she said, trying to get into my car one afternoon. I was taking her to McDonald's, the one out of three options for food within a thirty-minute drive of the motel. The other options were Denny's or the 7-Eleven, where all of us got most of our food. As she jumped into the car, she crushed her hand with her body and awkwardly caught her foot in the crook of the car door, falling over the stick shift into my lap. "Fuck, this arm is always hanging around like a shackle on me or something. Fuck if I know what to do with it!" But sometimes in soothing tones, she would say to her arm, "Tell me, I will give it to you, just tell me what you need."

She referred to her arm often as something set apart from her, sometimes even otherworldly. "This arm has a mind of its own. Sometimes I want to chop it off, really. I just want to take something big and sharp and—" With her whole body she made a chopping motion. "But I know it needs me," Betty said to me one late night after hooking. Her arm was set apart and yet a part of her, and she related to it as such. We'd had many conversations about the things we did as kids: eating paint, detergent, dirt, sand, paper. I told Betty about the first house I can remember living in on

Grant Street in Hartford. In this two-bedroom house, we somehow fit five and sometimes six adults (when my great aunt came for extended stays) and up to eight children, depending on who had stolen money from whom or who was on the verge of bankruptcy. She told me about growing up with eight siblings and the fights they would have and the "shit they would get into" with the other kids in the neighborhood. She talked about how "bois" were better than men and how she "ran around with [her first boi] when she was thirteen" and how "men cause so much strife."

We talked about living with chronic pain and how it changes your relationship to the world. How we had in common the experience of seizures, which we agreed were like windows into another world. Betty told me about how she could induce hers and how when she ran out of money, her seizures would help her engineer her way into a few days' rest. In the hospital, sometimes the doctors would do tests on her arm but give up when the answer didn't come easily or tell Betty that the problem was "in her mind." On occasion, Betty's arm would move not of her own volition and she would talk to it: "Calm down, would you! What the fuck do you want from me anyway?" One day I finally asked, "What *do* you think it wants?" Without flinching, Betty responded, "To go home, of course."

As the setting sun peeked through her partially open curtains, Betty sunk into her high, readying herself to tell me a story. "Baby girl, get your recorder." She sat with her head against the lacquered headboard, hair half cornrowed, half picked out. The pick still in her hair, every now and then she'd reach up and scratch her head with it. She shifted from one hip to another, trying to hide her pain, or perhaps this was just what it was always like for Betty to move in her body. She made a sharp noise as she lowered a hip back down to the motel bed. I'd seen her do this before when we would scatter from the parking lot, lifting ourselves from the low platforms of plastic chairs, turned-over motel garbage cans, tires from the side of the road. It wasn't just Betty; all the girls, myself included, had stiffness—aches and pains that would make their presence audible but that were not diagnosable.

"I come from everywhere," she began, "and I was born nowhere. I walked out of the swamp like some goddess creature covered in mud, with my tits and my big ol' butt hanging out." She laughed, then continued.

The white men's tongues hung out of their mouths like they ain't never seen anything like me before. They was hungry, so hungry, their dicks so hard they couldn't contain themselves, and I just walked around telling them to kneel in the dirt and mud. The men loved how I moved through it too, and they marveled at how my belly touched the water. They was scared too, because they knew I could make the swamp do things to them, swallow they ass whole. Most people don't know that this where I come from and where I disappear to, baby girl, but I'm telling you because I know you know that you come from there too. These white men know that my snatch could shoot tentacles out and that my ass could swallow them whole. They don't know how much they want me but they hungry for me. They don't understand that the swamp whispers sweet nothings in my ear. They don't know that I don't care about them because they kneel before me and I could crush them. I told them, "I'm a fucking swamp mermaid, King, GodDESS queen, and you will kneel in shit just to look upon me!"

At the edge of her motel bed, I gazed upon Betty as I had done many times before. The first night I saw her, she was leaning against the paint-chipped baby-blue double railing that rimmed the back of the motel. Tracks of cornrows led up the mountain of her perfectly coiffed beehive. She had one foot wrapped around one of the railing's perpendicular poles, and perched in her neon-pink sweatpants, she blew her cigarette smoke to the Virginia night sky, swinging her free foot back and forth. Bordering on orthopedic, her sneakers were prescient of the nostalgic return of early '90s oversized Black fashion. Her giant white T-shirt matched her big shiny white sneakers. Her three-and-a-half-inch bamboo hoops—which always sat more on her left shoulder than her right—spelled out, in cursive, "Tracee." With two *e*'s. She went by many names with me, but Tracee was never one of them. Her sneakers glistened not because they were new but because she cleaned them almost every day with commercial bathroom soap or motel shampoo or antibacterial hand sanitizer or even windshield wiper fluid that she would surreptitiously procure from the gas station with a scavenged soda bottle. These were the things that she wanted me to know. She wanted me to know that "she was fly as fuck."

And Betty was always watching, observing me too. Experienced in being overdetermined by the political imagination of Black abjection, she observed me: my height, my weight, my walk; then my hair, my shoes, my clothes, my hoops. I was inscrutable in the racial taxonomies of the South, so she wondered out loud, "Where you from, baby; what are you doing around these motels? You don't look like you belong." Later, Betty would suggest that the way I move—being Black, a woman, and *able* to travel long distances alone and without men—was a marker of queer Black belonging, but for now, on this night, it led to a different presumption. Betty asked, "You hookin'"?

Laughing, I said, "No, I'm just trying to find my way down here." I explained to Betty that I lived in New York City, that I was doing research in Virginia about long-distance trash management and racism. I said, "The city sends most of its trash down here, and I guess I'm just trying to find it . . . ?"

With a familiar understanding I had heard before from Black working-class women, Betty said, "Baby, you've found it! It's everywhere; there is trash all over these parts. This whole part of Virginia is built on swamp trash; this motel was probably built on swamp trash. Shit, we're all swamp trash!"

Her head hung lower as her high changed shape and I gazed into her swamp-filled horizon. After all, *Most people don't know that this where I come from and where I disappear to, baby girl, but I'm telling you because I know you know that* you come from there too. Betty was a *swamp mermaid, King,* GodDESS *queen.* A miraculous vision of alternative origin. An ecology so queer, distinctions between human and nonhuman, swamp and filth, were indistinguishable. To speak of cis-sexism or heteropatriarchy would be to impose upon her the very things that she recomposed, a way to Be Fly. A way to fly. I saw her rise out of the water, her naked body glistening with wet earth. I saw her with tentacles that whipped the air and the men who had tried so very hard to subject her to their fantasies of containment, of domination. I heard her make sounds that could shake the water, like she was birthed from it, emergent mud and flesh combined. She could shape-shift, bend things for joy. She could contort herself for her own pleasure. I gazed upon her fantastical flights through her high at the corner of her partially smiling mouth and smiled back at her barely audible sounds of contentment kissing the stale cigarette motel air. Monstrous indeed. An agent of her own making, she dared to be free. She dared to *be.*

BLACK GENRES

Sometimes a lie is a way to be more than an anti-Black world can contain. Sometimes the truth is that being "a container, a mold" is unspeakable.[26] Sometimes the monstrous relations born of such plasticity are (Afro)futuristic in their conjuring of present pasts and past futures. Sometimes these monstrous relations are indescribable mutilations that if/when announced, might just rearticulate a new condition of the Black subject's "arrival"—a spectacular scene of violence to be used as a future referent for Black becoming.[27]

As Fanon so famously said, "I came into this world imbued with the will to find a meaning in things. . . . I found that I was an object in the midst of other objects."[28] Within a Black ontology, to follow "The Thing" is to follow Blackness's capacity to know matter without scientific reason.[29] Blackening environments (elsewhere) transubstantiate flesh and the flesh of the biosphere simultaneously. And it is from elsewhere that we need to hear Black ecological critique. So much of this book has been an encounter with the practice of telling and retelling relationships to place and objects, to tell in ways that reveal how storied the material *always* is, and that there is no such thing as *things* with no history. It is also to do so as a Black *materialist* for whom the ongoing expropriation of native land and the dispossession of Black and Indigenous life inform an ontology that, as da Silva says, is a feminism that "has no qualms with thingness."[30] Moreover, I've tried to emphasize how histories themselves are lively in their material violences, disavowed by the authorized ways we tell/measure toxicity. As a result, telling, as I suggest at the end of chapter 4, might be a way of tending to our more-than-human world—a world in which Black living is an ecological struggle. Trash is a site to rethink a radical Black eco-imaginary that doesn't presume that the only appropriate relationship between people and nature is one that is pristine. In fact, it insists that not only do people come from places and places come from people, in the facts and fictions of an anti-Black world, sometimes the places named are not of this world.

The materialists in this book forward experiments with discards that chart an irreverence for property and its forms of reason. Unlike those materialists who disappear the structures of expropriation in favor of the agency of matter, discarded people know that discards have many capacities. For those discarded by the architecture that arranges the value of racialized life, the material world has always been a place to insist that there are other ways to be. As Denise Ferreira da Silva argues, the "Blackness . . . available to a Black Feminist Poethics . . . charts a terrain by . . . review[ing] its Categories, rearrang[ing] its project, and interrogat[ing] the very premises of its craft, without any guarantees that the craft itself will survive the exercise."[31] Simply put: Some things *can't be* expropriated, be it partying like a hood rat, Marty's *blues*, or the storied litter of gentrification that can't be sold even while hustling. The ecological struggle of Black life is also the struggle against the terms we give "the good life." Straining to tell these other stories is a way of refusing racial capitalism's names for *living* as clean and white; perhaps it is even a way of accessing lessons of a Black ontology that tell us what is required to survive the fact that waste is already shaping our future.[32]

To tell stories that *un*reason, to use misnames as a recycling practice, and to follow what da Silva argues is "Blackness's capacity to signify otherwise" is to tend to lives *anyway*, to notice how and what people steward with/in the conditions of alteration and how people improvise living room(s).[33] Elsewhere, people *know*, without the modern categories upon which whiteness's colonial authority rests. The condition of always being *excess* (to land) and a reservoir indexing property's ecological violence is to be confronted with the simultaneously imposed and fabulous capacity for transformation. Here, in the Black feminist elsewhere, "matter imaged as contingency and possibility rather than necessity and determinacy" might very well be the basis for a stewardship unhinged from *ownership*, living unhinged from *possession*, and cultivation unhinged from *capital*.[34]

In the ecological improvisations of discarded living, people tell different stories: about where they come from, how they're named, what they've seen, and how to testify. They speak environmentalisms that challenge the one we are most familiar with. Property's hold over our stories of, and entanglements with, matter is the environmentalism that gives the earth, the body, and the object one directive: to become propertied. Property is a *material* project with profound effects, and its violence impresses a directive that speaks its grammar loudly. Property (and its attendant methods of enclosure) demands that we read landscapes and the stories above for their extractability. But *things* are unruly, and they do not obey the directives they are given. Not everything can be possessed.

Notes

1 For more on the relationship between pesticides and colonial violence, see Agard-Jones, "What the Sands Remember"; Agard-Jones, "Bodies in the System"; Lyons, *Vital Decomposition*; Williams and Porter, "Cotton, Whiteness, and Other Poisons."

2 The 1986 Sexual Offences Act made it illegal for women to be lesbians. For more on a Black feminist critique of the act and its political implications, see M. J. Alexander, "Redrafting Morality." *Dougla* is a vexed racial term applied to those of mixed Indo-African heritage in Trinidad and Tobago. In "The Dougla in Trinidad's Consciousness," Ferne Louanne Regis (2011) argues that the term is etymologically related to its Indic origin, *dogla*, meaning "mutt" or "impure breed."

3 In "Mama's Baby, Papa's Maybe," Hortense Spillers opens with a long list of names for Black women in the national imaginary. These, she argues, are "markers so loaded with mythical prepossession that there is no easy way for the agents buried beneath them to come clean" (65). The welfare queen is a name among many for Black women's inherent degenerative capacities. See also Gregory, "Race, Rubbish, and Resistance," 27.

4 Gregory, *Black Corona*, 5–19; McKittrick, *Demonic Grounds*, ix–xxxi.

5 Finney, *Black Faces, White Spaces*, 34.

6 Outka, *Race and Nature*, 2.

7 Muhammad, *Condemnation of Blackness*, xxiii–xxvii.

8 New York City Mayor Rudolph Giuliani would produce his reputation as the man who "cleaned up" New York City based on this policing strategy. One of the major pillars of his cleanup strategy, which he again highlighted in his failed run for president in 2020, was the focus on policing what he called "quality-of-life crimes." These so-called quality-of-life crimes renamed Black place-making practices as antisocial and contagious behavior. W. J. Wright, "As Above, So Below"; T. Williams, "For 'Peace, Quiet and Respect'"; Mills, "Black Trash."

9 In *Demonic Grounds*, Katherine McKittrick argues that in order for geographic domination to render itself "natural," "obvious," and "absolute," it must enroll the material world of the dispossessed and subaltern

into the violence of colonial theft. The violence of spatial domination included the transatlantic kidnapping and "planting" of Black "human cargo," whose *labor* was central to carving up native territory—turning Turtle Island into the "obviously" or "naturally" white United States—but whose *living*, subject formation, and histories of resistance are rendered unplaceable or dis-locatable (x). The ungeographic, then, is a colonial projection, an operation that stabilizes colonial geographies and signals the way geography is less stable than white supremacy's management of space and methods of displacement (maps, jurisdiction, property) would make it seem. I take the "ungeographic" to be a logic that points us, paradoxically, to the fundamental spatiality of Black life; otherwise put, geography broadly is a contested "terrain of struggle" (xx).

10   King, *Black Shoals*, xi.

11   In "Territory as Analytic," Joanne Barker reminds us that when we center Indigenous territory as an analytic, we see how "dispossession is not normal, nor is it a thing of the past. It is not done—it is a *doing*" (31).

12   Wynter, "Unsettling the Coloniality of Being/Power/Truth/Freedom"; Ferdinand, *Decolonial Ecology*.

13   Trask, "The Color of Violence," 81.

14   Tuck and Yang, "Decolonization Is Not a Metaphor," 6–7.

15   Harris, "Whiteness as Property"; Liboiron, *Pollution Is Colonialism*, 132.

16   Barker, "Territory as Analytic," 31.

17   In *Development Arrested*, Clyde Woods theorizes the resilience of plantation relations throughout and after Reconstruction. His tracing of the way the planter class continuously reconstituted itself after the Civil War denaturalizes the relationship between development and the environments of white people in the south (4–16). I am also thinking here of Mishuana Goeman's "Disrupting a Settler-Colonial Grammar of Place," in which she theorizes "settler grammar" as an undergirding structure of settler relations to place (235–61). Taken together, property is not only a settler social relation but also a relationship predicated on ecological devastation that condemns Indigenous sovereignty and Black people's relationship to the material world.

18   Liboiron, *Pollution Is Colonialism*, 40.

19   W. J. Wright, "As Above, So Below," 794.

20   Da Silva, "Toward a Black Feminist Poethics," 92–93. I draw on this passage to emphasize da Silva's argument that the Black feminist poet has a special relationship to "the thing," a way of seeing matter and flesh as possibility. While da Silva does not argue that property is what fundamentally occludes that possibility, I see her assertation that obfuscating the "expropriation of the productive capacity of the conquered lands and enslaved bodies" is central to the ecological facts and fictions of *property* (83).

21 Murphy, "Alterlife and Decolonial Chemical Relations," 497.

22 Murphy, "Alterlife and Decolonial Chemical Relations," 497.

23 See Kelley, "'We Are Not What We Seem,'" 79–80.

24 Nir, "New York City Fights Scavengers over a Treasure."

25 Horton-Stallings, *Dirty South Manifesto*, 5.

26 Hernandez, *Aesthetics of Excess*, 11.

27 Armiero, *Wasteocene*, 54.

28 Armiero, *Wasteocene*, 54.

29 Armiero, *Wasteocene*, 49.

30 Ticktin, "Migrant Occupations."

31 Armiero, *Wasteocene*, 49.

32 See Roane, "Plotting the Black Commons."

33 Roane, *Dark Agoras*, 29–62.

34 Douglass, *Heroic Slave*, 17.

35 Diouf, *Slavery's Exiles*.

36 Bauman, *Wasted Lives*.

37 See Mills, "Black Trash."

38 Woods, *Development Arrested*.

39 Millar, *Reclaiming the Discarded*, 5 (my emphasis).

40 Strasser, *Waste and Want*.

41 Millar, *Reclaiming the Discarded*, 9.

42 Tadiar, *Remaindered Life*, 13, 6.

43 Millar, *Reclaiming the Discarded*, 9.

44 See Armiero, *Wasteocene*; Liboiron, *Pollution Is Colonialism*; Iheka, *African Ecomedia*; Murphy, "Alterlife and Decolonial Chemical Relations"; Frazier, "Troubling Ecology."

45 Rathje and Murphy, *Rubbish!*, 11.

46 Hird, "Knowing Waste," 455.

47 Stoler, *Imperial Debris*, 10.

48 Glissant, *Poetics of Relation*.

49 Robinson, *Black Marxism*; C. L. R. James, *Black Jacobins*; Marable, *How Capitalism Underdeveloped Black America*; Mbembe, *On the Postcolony*; Gilmore, *Golden Gulag*; Nixon, *Slow Violence*; Snorton and Haritaworn, "Trans Necropolitics."

50 In *Democracy's Infrastructure*, Antina von Schnitzler details the post-apartheid technopolitics of Johannesburg, South Africa. Post-apartheid water rationing became a way to mediate racialized people's relationship to the state, including by making it increasingly difficult to dispute that uneven access to water was a *result* of the state. While von Schnitzler does not argue that infrastructure is racist, I read this uneven techno-materialization of power as just one of many examples of how infrastructure relies on uneven access.

51 Osborne, "Security and Vitality."

52 Lea and Pholeros, "This Is Not a Pipe"; Mrázek, *Engineers of Happy Land*.

53 Hawkins, "Plastic Bags"; Hawkins, "Down the Drain"; Hawkins, *The Ethics of Waste*; Masco, *Nuclear Borderlands*; Reno, *Waste Away*.

54 Reno, *Waste Away*, 23 (my emphasis).

55 Nixon, *Slow Violence*.

56 Mills, "Black Trash," 84; Liboiron, *Pollution Is Colonialism*.

57 Bullard, *Dumping in Dixie*, 98; Pulido, "Rethinking Environmental Racism"; Pulido, "Flint, Environmental Racism and Racial Capitalism"; Davies, "Toxic Space and Time"; W. J. Wright, "As Above, So Below"; Liboiron, *Pollution Is Colonialism*; Agard-Jones, "Bodies in the System."

58 Ferdinand, *Decolonial Ecology*, 25–35.

59 Ferdinand, *Decolonial Ecology*, 9.

60 Harris, "Whiteness as Property," 1745–46.

61 Nichols, *Theft Is Property!* See also Wynter, "Unsettling the Coloniality of Being/Power/Truth/Freedom."

62 Wilderson, *Red, White and Black*, 207.

63 Wolfe, "Settler Colonialism and the Elimination of the Native"; Tuck and Yang, "Decolonization Is Not a Metaphor"; Snelgrove et al., "Unsettling Settler Colonialism."

64 Whyte, "Settler Colonialism, Ecology, and Environmental Injustice; A. Simpson, *Mohawk Interruptus*; Morgensen, *Spaces Between Us*; Rowe and Tuck, "Settler Colonialism and Cultural Studies." Though the settler was successful in producing and maintaining the violence of his own sovereignty, in "The State Is a Man," Audra Simpson reminds us that his successes were never (and will never be) complete.

65 As in John Locke's treatise on property, "The wild woods and uncultivated waste of America, left to nature without any improvement, tillage, or husbandry" (19), signals the need to transform nature to the highest order of property. Moreover, "Any one has liberty to make use of the waste" naming white men's burden to make use of untamed nature (61). "Land that is left wholly to nature, that hath no improvement of pasturage, tillage, or planting, is called, as indeed it is, waste" (20). For more on colonial discourse of waste and property, see Wolfe, "Settler Colonialism and the Elimination of the Native"; Hird and Zahara, "Arctic Wastes"; Stamatopoulou-Robbins, *Waste Siege*; Hird and Wilkes, "Colonial Ideologies of Waste"; Wideman, "Property, Waste, and the 'Unnecessary Hardship' of Land Use Planning."

66 Ahuja, "Intimate Atmospheres," 372.

67 McKittrick, "On Plantations, Prisons, and a Black Sense of Place."

68 Yusoff, *Billion Black Anthropocenes or None*; Schuller, *Biopolitics of Feeling*; Jackson, *Becoming Human*; Karera, "Blackness and the Pitfalls of Anthropocene Ethics."

69 Wynter, "Unsettling the Coloniality of Being/Power/Truth/Freedom";
    Weheliye, *Habeas Viscus*; Yusoff, *Billion Black Anthropocenes or None*;
    Nichols, *Theft Is Property!*

70 Liboiron and Lepawsky, *Discard Studies*.

71 Wolfe, "Settler Colonialism and the Elimination of the Native," 387–90.

72 Kelley, "'We Are Not What We Seem'"; Diouf, *Slavery's Exiles*; J. B. Morris,
    Dismal Freedom.

73 Tadiar, *Remaindered Life*, 13; Bledsoe, "Marronage as a Past and Present
    Geography"; W. J. Wright, "As Above, So Below."

74 Fanon, *Black Skin, White Masks*; Collins, "Gender, Black Feminism, and
    Black Political Economy"; hooks, *Feminism Is for Everybody*; K.-Y. Taylor,
    "Combahee River Collective Statement."

75 Browne, *Dark Matters*; D.-A. Davis, "Politics of Reproduction"; Roberts,
    Killing the Black Body; Strings, *Fearing the Black Body*.

76 Armiero, *Wasteocene*, 49.

77 Morgan, *Laboring Women*; Oyěwùmí, *Invention of Women*.

78 Messerschmidt, "'We Must Protect Our Southern Women,'" 78.

79 Collins, *Black Sexual Politics*.

80 Spillers, "Mama's Baby, Papa's Maybe," 75.

81 For Hortense Spillers's use of "ungendering," see Spillers, "Mama's Baby,
    Papa's Maybe." For a helpful look at the uses of "ungendering," see Pinto,
    "Black Feminist Literacies."

82 Spillers, "Mama's Baby, Papa's Maybe," 68; Snorton, *Black on Both Sides*,
    55–98.

83 Bey, *Cistem Failure*, 22.

84 Snorton, *Black on Both Sides*, 83.

85 C. J. Cohen, "Punks, Bulldaggers, and Welfare Queens," 441.

86 C. J. Cohen, "Deviance as Resistance," 38.

87 Snorton and Haritaworn, "Trans Necropolitics"; Williamson, *Scandalize
    My Name*.

88 Haley, *No Mercy Here*, 199.

89 Haley, *No Mercy Here*, 199.

90 Haley, *No Mercy Here*, 200.

91 Agard-Jones, "What the Sands Remember"; Roane, "Black Ecologies,
    Subaquatic Life"; W. J. Wright, "As Above, So Below."

92 Heynen and Ybarra, "On Abolition Ecologies"; Heynen, "Plantation Can
    Be a Commons."

93 Roane, "Plotting the Black Commons"; Roane, *Dark Agoras*; W. J. Wright,
    "As Above, So Below"; J. B. Morris, *Dismal Freedom*; Allewaert, *Ariel's
    Ecology*.

94 Campt, *Listening to Images*.

95 McKittrick, "Mathematics Black Life," 16.

96 Gilmore, *Golden Gulag*, 28.

97   J. James, "Architects of Abolitionism."

98   Allen, "Black/Queer/Diaspora at the Current Conjuncture."

99   Fabian, *Time and the Other*, 37.

100 Hartman, "Venus in Two Acts"; Williamson, *Scandalize My Name*.

101 Cervenak, *Wandering*, 22–23.

102 As Nadja Eisenberg-Guyot argues in "On How to Live While Being Thrown Away," people sustain pleasurable ways of life through breaking the law. While in their work the "crime" is drug use, their political insistence on pleasure reminds us that even our "liberal" critiques of economic constraint often pathologize poor people of color. Moreover, when pathology sneaks in, it forecloses the possibility that poor people of color who live on the street, who survive with waste, and who use drugs have robust political visions.

103 In her book *Against Purity*, Alexis Shotwell calls for a challenge and a way to think about living ethically in compromised times.

104 Horton-Stallings, *Dirty South Manifesto*.

FLOW

1   According to New York City census data, the population of Staten Island was 80 percent white in 1990 and about 73 percent white in the early 2000s. See Tumarkin and Bowles, "Staten Island"; New York City Department of City Planning, NYC 2000.

2   Archives of the Mayor's Press Office, "Mayor Giuliani and Borough President Molinari Mark Exportation of Residential Trash from the Bronx: Contract Sends Trash Out-of-State—Instead of to Fresh Kills—for the First Time in 60 Years," press release, July 1, 1997, https://www.nyc.gov/html/om/html/97/sp388-97.html.

3   Melosi, *Fresh Kills*.

4   The Triborough Bridge and Tunnel Authority also built highways through the Bronx, which are responsible for the excessive rates of asthma in the borough's Black and Puerto Rican residents.

5   Moses was planning for the siting of the world's fair, and the Rikers Island landfill, which would be visible from the fair, was an eyesore. Thus, he was going to need to relocate the landfill, and he chose the wetlands of Staten Island to do it. See Steinberg, *Gotham Unbound*.

6   Smith, "Giuliani Time," 1.

7   Lipton, "Five States Challenge New York's Trash Plan." The waste contract was the city's first step toward Giuliani's goal to export a total of 12,900 tons of residential and commercial waste per day. See Lipton and Melton, "Tons More of N.Y. Trash Headed for Va."

8   Lipton, "Five States Challenge New York's Trash Plan."

9   Lipton and Melton, "Tons More of N.Y. Trash Headed for Va."

10   Smith, "Giuliani Time," 1.

11   Smith, "Giuliani Time," 1.

12   Smith, "Giuliani Time," 3.

13   A project we still see today, most notably within the militarized police state of Mayor Eric Adams.

14   Smith, "Giuliani Time," 3.

15   For more on waste contracts, see Sze, *Noxious New York*, 109–41; Melosi, *Garbage in the Cities*; Melosi, *Sanitary City*.

16   Newman, "Giuliani's Trash-for-Culture Deal"; Melton, "N.Y. Mayor's Trash Talk Riles Va."

17   The Virginia governor's push in the general assembly was unsuccessful in large part due to the 1994 Supreme Court decision that protects waste's travel under the Constitution's Commerce Clause. See c & a Carbone, Inc. v. Clarkstown, 511 U.S. 383 (1994).

18   Ultimately, Jim Gilmore's grandstanding was a show to his constituents of his stated commitments to the environment, even as his campaign budgets revealed he had received over $100,000 in gifts and donations from the waste industry.

19   Waste Management Holdings, Inc. v. Gilmore, 87 F. Supp. 2d 536 (E.D. Va. 2000); Waste Management Holdings, Inc. v. Gilmore, 252 F.3d 316 (4th Cir. 2001). There were five statutory provisions signed into law under James Gilmore; all were deemed unconstitutional.

20   Deborah Cowen describes the ascent of logistics from a business "strategy" to a capitalist force and industry unto itself. As a surveilled and protected industry, logistics is primarily concerned with the *flow* of goods. When things threaten *flows*, such as borders, regulations, and, importantly, laborers, they are reclassified as "threats" to the health of the nation. See Cowen, *Deadly Life of Logistics*, 21–52.

21   Chahim, "Logistics of Waste."

22   See Roane, *Dark Agoras*.

23   Woods, *Development Arrested*, 4–12.

24   Roane, *Dark Agoras*, 68.

25   Roane, *Dark Agoras*, 68.

26   Mills, "Black Trash," 82.

27   Lipton and Melton, "Tons More of N.Y. Trash Headed for Va."

28   Roane, "Black Ecologies, Subaquatic Life."

29   In another context, Dean Chahim writes about the logistics of managing wastewater within the self-devouring development of Mexico City. Having built luxury apartment buildings on top of the reservoirs engineered to receive floodwaters during heavy rains, engineers are tasked with the death-dealing flows of wastewater. With no other options but to route the flows of wastewater through poor neighborhoods, engineers

have to knowingly make calculations about whose life is "valuable" within the self-devouring constraints of constant growth and development. Chahim, "Logistics of Waste."

30   Moreton-Robinson, *White Possessive*, x–xx, xix.

CHAPTER ONE. TOXIC CAPTURE

1   T. Williams, "For 'Peace, Quiet and Respect,'" 498.

2   T. Williams, "For 'Peace, Quiet and Respect,'" 498.

3   Fanon, *Wretched of the Earth*, 39.

4   Fanon, *Wretched of the Earth*, 38.

5   Particularly when it comes to environmental risk, public health literature demonstrates over and over that it is concentrated in poor neighborhoods of color. See *Toxic Wastes and Race in the United States*; Mohai and Bryant, "Environmental Injustice"; Zimmerman, "Social Equity and Environmental Risk"; Faber and Krieg, "Unequal Exposure to Ecological Hazards." From sickness on the plantation to redlining to predatory lending, financial risk is concentrated. For sickness on the plantation, see Bonhomme, "Contagion on the Plantation." Racialized punishment, condemnation, and predatory inclusion—alongside superexploitation and accumulation by dispossession—put the *racial* in racial capitalism and are central logics through which capital is produced, circulated, and hoarded.

6   See Armiero, *Wasteocene*; Vasudevan, "Intimate Inventory of Race and Waste."

7   Mills, "Black Trash."

8   Liboiron, *Pollution Is Colonialism*, 40.

9   Césaire, *Discourse on Colonialism*.

10   Hartman, *Scenes of Subjection*, 19.

11   Bullard, *Dumping in Dixie*.

12   Harney and Moten, *Undercommons*.

13   Browne, *Dark Matters*, 31–33.

14   See also W. J. Wright, "As Above, So Below."

15   Finney, *Black Faces, White Spaces*.

16   Karera, "Blackness and the Pitfalls of Anthropocene Ethics," 45–46.

17   W. J. Wright, "As Above, So Below," 796.

18   The hip-hop duo comprising No Malice and Pusha T is from Virginia Beach. Their music often describes the history of Southern racism and its relationship to drugs, poverty, and pleasure in the Tidewater.

19   In "*Loving v. Virginia* as a Civil Rights Decision," Dorothy Roberts argues that when the Supreme Court struck down miscegenation laws, the court had all the "scientific" evidence that it needed to find race a social,

not biological, category. Had the Supreme Court decided to "refute the validity of race as a biological category," it could have been "an enduring strike," as opinion writer Brent Staples writes in the *New York Times*, "against white supremacy in the U.S." Staples, "What If the Court in the Loving Case Had Declared Race a False Idea?"

20 My use of "girls" here is a term of endearment. "The girls" referred to themselves as such, so I do too. I take a moment to note this because of how Black people in the United States have been subject to gendered violence, including in the use of the diminutive term *boy* for Black men. Similarly, *girl* has been used to undermine Black women as political actors, a status presumed to emanate from masculinity. Even as Black women are diminished by the term *girl*, actual Black girls are denied childhood, whether through their hypersexualization and exposure to sexual violence, or through the state's pursuing, prosecuting, and murdering Black girls, never afforded the innocence of childhood. For more on Black girlhood, see Williamson, *Scandalize My Name*; Field and Simmons, *Global History of Black Girlhood*.

21 Latour, *We Have Never Been Modern*; B. Brown, "Thing Theory"; Barad, "Posthumanist Performativity"; Braidotti, *Metamorphoses*; Jane Bennett, *Vibrant Matter*.

22 See Visperas, *Skin Theory*; Sexton, "Social Life of Social Death."

23 Ferdinand, *Decolonial Ecology*.

24 See Urban Renewal Center, "Changing Tides." For more scholarly references, see F. R. White, *Black, White and Brown*; Parramore, *Norfolk*. For more on the local reporting, see Turken, "'It Was Horrific.'" For more on the history of the FHA and redlining more broadly, see Rothstein, *Color of Law*.

25 Cox, "Naval Power Plant Proposal Tests Virginia on Environmental Justice."

26 E. Simpson, "Lead Poisoning."

27 Silkenat, *Scars on the Land*. EPA and city documentation of the Suffolk landfill also refer to the site as the Hosier Road landfill, which is now a Superfund site.

28 In addition, new EPA regulations for landfills made the cost of retrofitting inaccessible to poor cities and many rural towns.

29 Environmental Protection Agency, *Fourth Five-Year Review Report for Suffolk City Landfill*, 3.

30 These numbers vary slightly depending on the report. An initial health assessment conducted by the US Department of Energy claimed twenty-seven tons, while the initial remedial investigation report cited thirty tons. See US Department of Energy, "Health Assessment for Suffolk City Landfill; SCS Engineers, *Remedial Investigation Report for the Hosier Road Landfill*.

31 The National Priorities List refers to the EPA's Superfund sites priori-
    tized for remediation.

32 McHugh, "Landfill Added to EPA Superfund List."

33 SCS Engineers, *Remedial Investigation Report for the Hosier Road Landfill*,
    1–17, 3–18.

34 In *Waste Siege*, Sophia Stamatopoulou-Robbins argues that the waste
    siege on Palestinian lives produces material effects.

35 In "Obese Black Women as 'Social Dead Weight,'" Sabrina Strings writes
    of the contradictory nature between research on obesity and the way
    Black women are constructed as "deadly agents of disease" (108). This
    structure of blame is also explored by Dorothy Roberts in *Killing the Black
    Body*, in which Roberts traces this relationship to reproductive coercion
    and health care in the United States. For more on Black women's health,
    see Strings, *Fearing the Black Body*; Fullwiley, *Enculterated Gene*; D.-A.
    Davis, *Reproductive Injustice*; Schalk, *Black Disability Politics*.

36 Coren, "50 Years of Research."

37 In 1976, Herbert Needleman's research on the effects of lead on childhood
    development showed that even low levels cause "central nervous system
    impairment." He was subsequently attacked, including for advocating
    for exposure prevention protocols that continue to undergird debates
    about how much the state should act. See Hernberg, "Lead Poisoning in
    a Historical Perspective," 250.

38 Centers for Disease Control and Prevention, "Recommended Actions
    Based on Blood Lead Level."

39 Gilbert and Weiss, "Rationale for Lowering the Blood Lead Action Level."

40 Murphy, "Chemical Regimes of Living," 700–701.

41 See also Vasudevan, "Intimate Inventory of Race and Waste"; Sharpe, *In
    the Wake*.

42 Hare, "Black Ecology."

43 Bullard, "Environmental Justice in the 21st Century."

44 McKittrick, *Dear Science and Other Stories*, 3.

45 McKittrick, *Dear Science and Other Stories*, 3.

46 McKittrick, *Demonic Grounds*, xii. See also Hawthorne, "Black Matters
    Are Spatial Matters."

47 I use Cheryl Harris's term "vested interest in whiteness" to emphasize
    white supremacy's entanglement with law and regulation. Harris,
    "Whiteness as Property," 1725.

48 "SPSA Landfill Expansion in Jeopardy," *Virginian-Pilot*, March 3, 2008.

49 WTE facilities, thus, tend to authorize and compound the vulnerability
    of people and land already vulnerable to risks of multiple kinds.

50 Sassen, *Global City*, 110–22.

51 DeSouza et al., "Nationwide Study of Short-Term Exposure to Fine
    Particulate Matter"; Allen et al., "Fine Particulate Matter Air Pollution."

52 Virginia Department of Health, *Virginia Asthma Plan, 2011–2016*, 2.

53 Virginia Department of Health, *Virginia Asthma Plan, 2011–2016*, 3.

54 The *Virginia Asthma Burden Report*, produced in 2018, used data from the Virginia Department of Health's Behavioral Risk Factor Surveillance Survey from 2011 to 2016. It is worth pointing out that this data was collected from the same five years in which the Virginia Asthma Coalition's plan was to *lower* asthma rates. The *Asthma Burden Report*, however, reveals that the coalition's plan made no measurable impact.

55 After a number of environmental problems, including a fire in Portsmouth, Virginia, in 2019 and the company's facility in Millbury, Massachusetts, being named one of the top pollutant emitters in the country the same year, Wheelabrator—once a subsidiary of WMI—rebranded as WIN Waste Innovations. While the research conducted for this book was done prior to Wheelabrator's dirty reputation (and new name), it's also worth noting, as Cyrus Moulton writes, "Wheelabrator Millbury may be dirtier than many of its brethren [but], it is generally meeting emissions limits established by the Department of Environmental Protection and the U.S. Environmental Protection Agency." Moulton, "Wheelabrator Millbury Is Cited as a Top Polluter."

56 Cox, "Naval Power Plant Proposal Tests Virginia on Environmental Justice."

57 Cox, "Naval Power Plant Proposal Tests Virginia on Environmental Justice."

58 Cox, "Naval Power Plant Proposal Tests Virginia on Environmental Justice."

59 Cox, "Naval Power Plant Proposal Tests Virginia on Environmental Justice"; Environmental Protection Agency, "Superfund: National Priorities List."

60 Cox, "Naval Power Plant Proposal Tests Virginia on Environmental Justice."

61 Environmental Protection Agency, "Superfund: National Priorities List."

62 The lone apposing vote came from Hope F. Cupit, one of two Black women on the board. It should come as no surprise that the opposing vote came from someone with a history of working on issues of water and housing in the urban and rural South.

63 Cowen, *Deadly Life of Logistics*, 62–68.

64 Harper, "SPSA Says Chesapeake Suit Is Causing Financial Woe."

65 The church has struggled to claim its status as a historical landmark for a variety of reasons, including that in the 1960s insects and soil-borne termites had eaten through many of the church's pews, handmade by the formerly enslaved. The National Register of Historic Places, too, cited a missing belfry for denying the church's historical legacy.

66 Sze, *Noxious New York*, 116–40.

67 That same year, Chesapeake tried to get out of its membership con-
    tract. Though unsuccessful, it was yet another critique leveraged at the
    regional cooperative waste service.

68 As a refuge from plantation violence, the Great Dismal Swamp was
    home to the largest maroon population in the United States and was a
    critical node in the fugitive map known as the Underground Railroad.

69 Wilder, *Covenant with Color*; Osman, *Invention of Brownstone Brooklyn*.

70 Reese, *Black Food Geographies*, 8.

71 Mitman, *Breathing Space*, 134.

72 Bruce, *How to Go Mad Without Losing Your Mind*, 2.

73 W. J. Wright, "As Above, So Below," 792, 797.

74 K. Simmons, "Settler Atmospherics"; Sharpe, *In the Wake*.

75 See Browne, *Dark Matters*; W. J. Wright, "As Above, So Below."

76 Under white supremacist heteropatriarchy, the imagined "purity" of ecol-
    ogy coheres against the imagined chaotic, disordered, and pathological
    environments of Black people. Describing where Black people live as "de-
    graded," as opposed to a place produced by the relations of punishment,
    is a way of pushing back against the idea that Blackness *brings* degrada-
    tion, a problem that disrupts but is always outside the ecological.

77 Here I'm drawing on C. Riley Snorton's argument in *Black on Both Sides*
    that fungibility and fugitivity are mutually enveloping (67–81); Tiffany
    Lethabo King also cites this argument in *The Black Shoals* by tracking the
    furtive movements of Black and Indigenous resistance (24–25).

78 Fanon, *Wretched of the Earth*, 39.

79 Bruce, *How to Go Mad Without Losing Your Mind*, 2.

INFRASTRUCTURE

1 Peisner, "Va. Trash Imports Rise."

2 Da Silva, "Towards a Black Feminist Poethics," 83; McKittrick, "On Plantations, Prisons, and a Black Sense of Place," 949.

3 McKittrick, "On Plantations, Prisons, and a Black Sense of Place," 951.

4 Roane, "Plotting the Black Commons," 247.

5 Pulido, "Rethinking Environmental Racism," 533.

6 Pulido, "Rethinking Environmental Racism," 544.

7 In Malcom Ferdinand's *Decolonial Ecology*, he argues that a particular
   form of habitation, which he terms "colonial habitation," impedes our
   ability to see the ecological world. For him, "world" is an important inter-
   ruption to a colonial project of habitation that sees environments and
   ecological relations as disconnected, including how colonial habitation
   informs the kinds of environmental politics that emanate from conservation projects uninformed by Indigenous land relations the world over.

8   Liboiron, *Pollution Is Colonialism*, 40.

9   Joshua Bennett, *Being Property Once Myself*.

10 Gilmore, *Golden Gulag*; Gilmore, *Abolition Geography*; Davis, *Are Prisons Obsolete?*

11 Kaba, *We Do This 'til We Free Us*," 27–30.

12 Schept, "(Un)seeing Like a Prison."

13 Spade, *Normal Life*; Davis et al., *Abolition. Feminism. Now.* Kaba, *We Do This 'til We Free Us*; Whalley and Hackett, "Carceral Feminisms"; Bernstein, "Militarized Humanitarianism Meets Carceral Feminism."

14 Kaba, *We Do This 'til We Free Us*.

15 From unchallenged desires for more police; to an ever-expanding security apparatus; to laws and ideologies that indoctrinate a belief in the ontological truth of "the criminal"; to the spatialized race relations of white flight that decimate tax bases in poor neighborhoods, to which the prison becomes an economic fix, the prison is an infrastructure socialized by a racist, classist, cis-sexist heteropatriarchal imagination.

16 Ferdinand, *Decolonial Ecology*, 1–74.

17 Here I'm referring to Ruthie Gilmore's analysis of the 1990s prison boom in California, where the prison did not fix the problem of crime but rather served as an institution to "fix" the problem of surplus land. See Gilmore, *Golden Gulag*.

18 Reno, *Waste Away*, 216.

19 Benjamin, *Race After Technology*.

20 Hartman, *Lose Your Mother*, 6; Gilmore, *Golden Gulag*, 28.

CHAPTER TWO. BECOMING FILL

1   See Morrison, *Playing in the Dark*; Gregory, *Black Corona*; Summers, *Black in Place*; Hartman, *Wayward Lives*; McKittrick, *Demonic Grounds*; Roane, *Dark Agoras*; Kelley, *Freedom Dreams*.

2   O. E. Butler, *Kindred*.

3   See Grande, *Red Pedagogy*; Tuck and Yang, "Decolonization Is Not a Metaphor."

4   Du Bois, *Black Reconstruction*.

5   Tuck and Yang, "Decolonization Is Not a Metaphor," 6.

6   These areas are often referred to as high-risk zip codes, which seem to ideologically map criminality.

7   Shaw and Waterstone, "Planet of Surplus Life," 1788.

8   Marx, *Capital: An Abridged Edition*, 298; Shaw and Waterstone, "Planet of Surplus Life," 1788. To make this argument, Shaw and Waterstone draw on Fuentes, "'Garbage of Society,'" and Sassen, "Expulsions."

9   Tadiar, *Remaindered Life*.

10 Tadiar, *Remaindered Life*, 6.

11 Wynter, "Unsettling the Coloniality of Being/Power/Truth/Freedom"; Jackson, *Becoming Human*; Weheliye, *Habeas Viscus*. And living itself, Neferti Tadiar argues in *Remaindered Life*, is a form of labor (13).

12 In *Black Reconstruction in America*, W. E. B. Du Bois theorizes the heroic resistance of the slave as the withdrawal of enslaved labor to cross enemy lines (55–84). But as Black feminists have pointed out, "The general strike" was not only the withdrawal of enslaved men's labor but the *insurgent* withdrawal and socially reproductive labor of enslaved women. See Weinbaum, "Gendering the General Strike." See also Haley, *No Mercy Here*.

13 Bullard, "Environmental Blackmail in Minority Communities," 83.

14 Though it is beyond the purview of this book, toxic industries in the 1900s offered Black workers the "promise" of relief from sharecropping. Unknowingly bargaining their bodies, families, and futures, toxicity would come to mark the punishment associated with Black life working off the plantation. For more on toxicity and regimes of labor, see Vasudevan, "Intimate Inventory of Race and Waste." For more on race and "dirty" labor, see Zimring, *Clean and White*.

15 In particular, the super-exploitation of Black women. See Jones, *End to the Neglect*.

16 Bullard, "Environmental Blackmail in Minority Communities," 83.

17 Vasudevan, "Intimate Inventory of Race and Waste," 771.

18 Roane, "Black Ecologies, Subaquatic Life," 231.

19 Roane, "Black Ecologies, Subaquatic Life," 232.

20 Marable, *How Capitalism Underdeveloped Black America*.

21 There are many ways in which this is reinforced, from the ability to own property to the ability to move out of a toxic place.

22 Shaw and Waterstone, "Planet of Surplus Life," 1788.

23 Tuck and Yang, "Decolonization Is Not a Metaphor," 1.

24 Tuck and Yang, "Decolonization Is Not a Metaphor," 6.

25 Burden-Stelly, "Modern U.S. Racial Capitalism," 10.

26 Shaw and Waterstone, "Planet of Surplus Life," 1789, 1788.

27 Burden-Stelly, "Modern U.S. Racial Capitalism," 10.

28 Tadiar, *Remaindered Life*; Millar, *Reclaiming the Discarded*; McKittrick, "Mathematics Black Life."

29 See Heynen and Ybarra, "On Abolition Ecologies."

30 Agard-Jones, "What the Sands Remember"; Silkenat, *Scars on the Land*.

31 Tso, *Physiology and Biochemistry of Tobacco Plants*, quoted in Lukezic, "Soils and Settlement Location," 4.

32 Tso, *Physiology and Biochemistry of Tobacco Plants*, quoted in Lukezic, "Soils and Settlement Location," 4.

33 Silkenat, *Scars on the Land*, 6.

34 Silkenat, *Scars on the Land*, 7.

35 Silkenat, *Scars on the Land*, 2.

36 Williams and Porter, "Cotton, Whiteness, and Other Poisons," 499.

37 Agard-Jones, "What the Sands Remember," 333.

38 Pilcher et al., "Dream Home Nightmares."

39 Tuck and Yang, "Decolonization Is Not a Metaphor," 6.

40 McCabe, "Have a Sinking Feeling About Your Home?"

41 Uteuova, "After Slavery, Oystering Offered a Lifeline."

42 Uteuova, "After Slavery, Oystering Offered a Lifeline."

43 Uteuova, "After Slavery, Oystering Offered a Lifeline."

44 Fanon, *Wretched of the Earth*; Agard-Jones, "What the Sands Remember"; Agard-Jones, "Bodies in the System."

45 McCabe, "Have a Sinking Feeling About Your Home?"

46 US Department of Agriculture, *Soil Survey of City of Suffolk, Virginia*. From 1996 to 2000, across the Southern cities of the Hampton Roads that border the swamp's porous edges, soil became a public policy issue, not through the soil itself but through the building codes used by developers that did not take the soil into account. Foundational problems began to emerge as a topic of concern when a flood of homeowners started to complain about the cracks around their doors and windows, and a "sinking feeling" that they did not know what to attribute it to. See McCabe, "Have a Sinking Feeling About Your Home?"

47 Uteuova, "After Slavery, Oystering Offered a Lifeline."

48 Williams and Porter, "Cotton, Whiteness, and Other Poisons," 501.

49 K.-Y. Taylor, *Race for Profit*, 123–45.

50 Virginia Department of Transportation, "History of Roads."

51 Tipping fees, which are the price that landfills charge a municipality or company to dispose of waste on its premises, are one of the ways that waste companies cover the cost of their daily operations. For private mega landfills, these fees can be kept low because the sheer volume of contracts they hold offsets the cost of daily maintenance.

52 Environmental Protection Agency, "Flow Control and Municipal Solid Waste." Ostensibly a provision intended to protect the environment as a public good, flow controls actually determine *where* competition happens.

53 C & A Carbone, Inc. v. Clarkstown, 511 U.S. 383 (1994).

54 Harper, "SPSA Landfill Expansion in Jeopardy."

55 Rosengren, "Update."

56 Lynch, "State Warns Waste Authority About Leaks at Suffolk Landfill."

57 Murphy, "Chemical Regimes of Living," 697.

58 Shaw and Waterstone, "Planet of Surplus Life," 1789.

59 According to the 2015 annual survey report on Suffolk's regional solid waste disposal system, SPSA owns eight hundred acres of land, four hundred of which are empty.

60　Though landfills are required to go through a series of bureaucratic red
　　tape, they often begin their expansion plans without permission. Rarely
　　are expansion plans denied, and when they are, work-arounds are regu-
　　larly created to ensure that landfills expand.

61　Harper, "SPSA Landfill Expansion in Jeopardy."

62　Harper, "SPSA Landfill Expansion in Jeopardy."

63　Shaw and Waterstone, "Planet of Surplus Life," 1789.

64　Although granted a conditional use permit (CUP) to build cell 7, as of
　　2022, SPSA is still in the environmental impact scoping process, where
　　the environmental impact statement must meet the requirements of the
　　National Environmental Policy Act and the Council on Environmental
　　Quality regulations. Thus, while a CUP has been granted, the impacts of
　　cell construction are still being assessed.

65　Harper, "SPSA Landfill Expansion in Jeopardy."

66　There have been a number of new cases brought against the Hampton
　　Roads Sanitation District by Black oystermen. However, each case
　　continues to fail on the basis of property law: The state has the right to
　　pollute its "own" property, always creating the toxic capture of nonwhite
　　(and, in this case, marine) life. For more on these Blackened ecologies
　　and subaquatic life, see Roane, "Black Ecologies, Subaquatic Life."

67　Shaw and Waterstone, "Planet of Surplus Life," 1789.

68　Most obviously, we can see this manifest itself through predatory
　　lending. As Keeanga-Yamahtta Taylor writes in *Race for Profit*, "Race and
　　risk to property value" live on in the housing market through "enduring
　　racist assumptions about Black hygiene" and "white property owners
　　protecting their investments" (259).

69　This chilling presentation of working to death reminds us that Black
　　ecologies connect regimes of Black labor to the material arrangement of
　　all life. See McKittrick, "Plantation Futures," 2 (emphasis added).

70　See McKittrick, "Plantation Futures."

71　Harney and Moten, *Undercommons*, 38.

72　Burden-Stelly, "Modern U.S. Racial Capitalism," 10.

73　Hartman, *Lose Your Mother*, 6.

74　Butt, *Life Beyond Waste*.

75　Grove, *Green Imperialism*.

76　W. Byrd, "History of the Dividing Line," 23, 22.

77　W. Byrd, "History of the Dividing Line," 22.

78　W. Byrd, "History of the Dividing Line," 119, 36, 96.

79　W. Byrd, "History of the Dividing Line," 20.

80　Despite the twenty-three times the word *plantation* or sixteen times the
　　word *industry*—which also refers to the plantation—emerges in the text,
　　the word *negro* appears twice, the word *slavery* appears once, and the
　　word *slave* does not appear at all.

81 King, *Black Shoals*, 78.

82 The capitalization of "White" here is in direct reference to Tiffany King's argument about the necessary triad of violence that allows the settler to cohere. See King, *Black Shoals*, 75–77.

83 King, *Black Shoals*, 16–21.

84 W. Byrd, "History of the Dividing Line," 37.

85 W. Byrd, "History of the Dividing Line," 112.

86 Locke, *Second Treatise of Government*, 131–39.

87 Locke, *Second Treatise of Government*, 181, 264.

88 Harris, "Whiteness as Property."

89 Tuck and Yang, "Decolonization Is Not a Metaphor," 6.

90 Heynen and Ybarra, "On Abolition Ecologies," 29.

91 Douglass, *Heroic Slave*, quoted in Heyen and Ybarra, "On Abolition Ecologies," 28.

92 Roane, "Plotting the Black Commons."

93 Roane, "Plotting the Black Commons," 251.

94 Diouf, *Slavery's Exiles*, 209–29.

95 Heynen and Ybarra, "On Abolition Ecologies," 28.

96 Heynen and Ybarra, "On Abolition Ecologies," 26.

97 Heynen and Ybarra, "On Abolition Ecologies," 28.

98 The Chowanoke and the Powhatan are stewards of the Nansemond.

99 Originally the Dismal Swamp Company was chartered to extract trees for logging and draining the swamp for "farm" land, a suggestion that came from William Byrd after charting the dividing line between Virginia and North Carolina. The company's inability to turn the swamp into plantation lands eventually led the company to change course. In 1787, the Virginia assembly chartered the Dismal Swamp Canal Company, allowing the speculative land project to turn its resources toward building a canal through the swamp. It should be noted that not only was the canal dug by enslaved labor, but it did not yield the waterway or profits that the company had hoped it would. Over and over in the story of white supremacy's attempt to capture "potential" in the swamp, the swamp refused to yield to colonists' demands.

100 And the company would change from a land speculation company to a timber company and finally to a canal company, which lasted up until the Civil War.

101 Diouf, *Slavery's Exiles*, 209–29; J. B. Morris, *Dismal Freedom*.

102 The company failed to produce any real value until 1810 from lumbering. In the George Washington Papers at the Library of Congress, the Dismal Swamp Company's documents detail the controversies around shareholders' lack of profit. See "Resolutions of the Dismal Swamp Company, 1 May 1785."

103 Sayers, *Desolate Place for a Defiant People*; Grant, "Deep in the Swamps."

104 The fire had also been monitored by NASA. See Hutchins, "Official Declare Dismal Swamp Fire Out After 111 days."

105 Speiran and Wurster, *Hydrology and Water Quality of the Great Dismal
Swamp*, 1.

106 Speiran and Wurster, *Hydrology and Water Quality of the Great Dismal
Swamp*, 1.

107 US Fish and Wildlife Service, "Resistance and Refuge."

108 Indigenous refusal to trade with settlers in 1608 eventually led to the first
Anglo-Powhatan War (1609–14). Subject to decades of violence, the Nansemond were displaced from their ancestral lands along the Nansemond
River. After the third Anglo-Powhatan War (1644–46) the Nansemond
were "given land" on the "northwest and south Branches of the Nansemond River." See "Tribal History," Nansemond Indian Nation, accessed
November 17, 2022, https://nansemond.gov/tribal-history/.

109 "Nansemond Indian Nation," *Encyclopedia Virginia*.

110 Tuck and Yang, "Decolonization Is Not a Metaphor," 11; "Nansemond
Indian Nation," *Encyclopedia Virginia*.

111 Appadurai, *Social Life of Things*.

SURPLUS

1 MacBride, *Recycling Reconsidered*, 2.

2 This Greenpeace report, titled *Circular Claims Fall Flat Again*, was initially
published in 2018 and updated in 2022 with the same conclusion.

3 *Circular Claims Fall Flat Again*, 4.

4 This is one of the critical aspects of the Greenpeace report. The first
survey of MRFs to document rates of recycling (2020) found this
trend across the country. In the update to the report in 2022, this was
confirmed, and the trend seems to be growing because "there is no end-
market buyer." *Circular Claims Fall Flat Again*, 4.

5 MacBride, *Recycling Reconsidered*, 24–29.

6 Lepawsky, "Changing Geography of Global Trade in Electronic Discards"; Lepawsky and McNabb, "Mapping International Flows of Electronic Waste."

7 Reddy, "Reimagining E-Waste Circuits," 58.

8 Armiero, "Case for the Wasteocene," 426.

9 Armiero, "Case for the Wasteocene," 425.

10 Armiero and De Angelis, "Anthropocene," 348.

11 Müller, "Toxic Commons," 444.

12 See Gidwani and Maringanti, "Waste-Value Dialectic"; Kennedy, *Ontology of Trash*; Bataille, *Visions of Excess*.

13 Gidwani and Maringanti, "Waste-Value Dialectic," 116.

14   Marx and Engels, *Manifesto of the Communist Party*, 24.
15   Gidwani and Maringanti, "Waste-Value Dialectic," 117.
16   Gidwani and Maringanti, "Waste-Value Dialectic," 117 (my emphasis).
17   Environmental Protection Agency, "Flow Control and Municipal Solid Waste."
18   King et al., "Analysis of the Economics of Prison Siting."
19   Sze, *Noxious New York*.
20   Brenner and Theodore, "Neoliberalism and Urban Condition," 102.
21   Thomson, *Garbage In, Garbage Out*. "Though there are instances when minorities invite hazardous facilities into their communities, this is the exception rather than the rule. This practice does not account for the thousands of hazardous facilities that are sited in minority communities around the country." See D. E. Taylor, *Toxic Communities*, 145.
22   Gidwani and Maringanti, "Waste-Value Dialectic," 118.
23   Gidwani and Maringanti, "Waste-Value Dialectic," 118.

CHAPTER THREE. REVISIONS FROM ELSEWHERE

1   Shabazz, *Spatializing Blackness*, 21.
2   Shaw and Waterstone, "Planet of Surplus Life," 1791. See also Davies, "Slow Violence and Toxic Geographies."
3   Hartman, *Scenes of Subjection*, 191–206; W. J. Wright, "As Above, So Below," 792.
4   In "Geotheorizing Black/Land," the authors argue that the "ungeographic" describes power but does not describe *a Black relation to land*. As I read their intervention, the "ungeographic" makes *hearing* Black land relations a political struggle, but it does not make clear how Black people think about, talk about, and relate to a sense of being with or becoming with land. See Tuck et al., "Geotheorizing Black/Land."
5   Tuck and Yang, "Decolonization Is Not a Metaphor," 1, 6.
6   Fanon, *Black Skin, White Masks*; Sexton, "Social Life of Social Death"; Warren, *Ontological Terror*; McMillan, *Embodied Avatars*; Davis, *Women, Race and Class*; Spillers, "Mama's Baby, Papa's Maybe"; Hartman, *Scenes of Subjection*; Lorde, *Sister Outsider*; Haley, *No Mercy Here*.
7   Tuck and Yang, "Decolonization Is Not a Metaphor," 6.
8   Puar, *Right to Maim*; Schalk, *Black Disability Politics*, 10; Murphy, "Alterlife and Decolonial Chemical Relations," 497.
9   Cf. Marable, *How Capitalism Underdeveloped Black America*; Haley, *No Mercy Here*; Robinson, *Black Marxism*; Hartman, *Wayward Lives, Beautiful Experiments*.
10   W. J. Wright, "As Above, So Below," 797.
11   Shaw and Waterstone, "Planet of Surplus Life," 1787.

12 Douglas, *Purity and Danger*.

13 Here I'm also thinking about Patricia Hill Collins's work on the way *freak* imposes a colonial gaze that shapes the consumption of Black women and their sexual representations. See Collins, *Black Sexual Politics*, 119–48. I'm also thinking about the way Elana Resnick critiques the concept of "resilience," a description applied to marginalized communities such as the Roma, excavating the way environmental racism reveals itself in places not necessarily or explicitly "marked [for] toxic extraction and contamination." See Resnick, "Limits of Resilience," 224.

14 Finney, *Black Faces, White Spaces*, 32–50.

15 In *Imperial Intimacies*, Hazel Carby thinks through the way that her mixedness is a product of the violent histories that brought African, Indigenous, and British colonists together. This is an important and ongoing intervention into the conversation about being "mixed race." As I see it, Carby insists that we think about the historical conditions that make racial "mixes" possible and what it might mean to be caught by these sociopolitical conditions.

16 TallBear, *Native American DNA*.

17 Gidwani and Maringanti, "Waste-Value Dialectic"; Prasad, "Towards Dalit Ecologies."

18 hooks, *Feminist Theory*, 26.

19 Hosbey and Roane, "Totally Different Form of Living"; W. J. Wright, "The Morphology of Marronage."

20 Wynter, "Unsettling the Coloniality of Being/Power/Truth/Freedom," 300–301.

21 M'charek, "Salty," 18.

22 G. Lewis, "Once More with My Sistren," 4.

23 Wynter, "Unsettling the Coloniality of Being/Power/Truth/Freedom"; Frazier, "Troubling Ecology"; Karera, "Blackness and the Pitfalls of Anthropocene Ethics."

24 Wynter, "Unsettling the Coloniality of Being/Power/Truth/Freedom"; Tadiar, *Remaindered Life*.

25 Snorton, *Black on Both Sides*, 83; Bey, *Cistem Failure*.

26 Horton-Stallings, *Dirty South Manifesto*, 4.

27 Chowdhry and Beeman, "Situating Colonialism, Race, and Punishment"; Collins, *Black Feminist Thought*; Collins, *Intersectionality as Critical Social Theory*; Collins, *Lethal Intersections*; hooks, *Teaching to Transgress*; hooks, *Black Looks*; Lorde, *Sister Outsider*; Moraga and Anzaldúa, *This Bridge Called My Back*.

28 Sedgwick, "Paranoid Reading and Reparative Reading," 123–51; hooks, *Black Looks*; Sandoval and Davis, *Methodology of the Oppressed*; C. J. Cohen, "Deviance as Resistance"; McKittrick, *Demonic Grounds*; Cervenak, *Wandering*; Cervenak, *Black Gathering*; McMillan, *Embodied Avatars*;

Haley, *No Mercy Here*; Williamson, *Scandalize My Name*; Hartman, *Wayward Lives*.

29 Hall, "Race, Articulation, and Societies Structured in Dominance"; Woods, *Development Arrested*; Roane, *Dark Agoras*; W. J. Wright, "As Above, So Below"; McMillan, *Embodied Avatars*.

30 Crenshaw, "Beyond Racism and Misogyny"; hooks, "Sexism and Misogyny"; Gilroy, *Black Atlantic*; Adams and Fuller, "Words Have Changed but the Ideology Remains the Same"; Hunter and Soto, "Women of Color in Hip Hop"; Pough, "What It Do, Shorty?"; Durham et al., "Stage Hip-Hop Feminism Built"; Snorton, "Referential Sights and Slights"; J. Brown, "Hip Hop, Pleasure, and Its Fulfillment"; Nyong'o, *Afro-Fabulations*; T. R. White, "Missy 'Misdemeanor' Elliott and Nicki Minaj"; Summers, *Black in Place*; Denise, "Where House Found a Home"; Denise, "Dancing Between Worlds"; Snorton, "Referential Sights and Slights"; Horton-Stallings, *Dirty South Manifesto*; Shabazz, "'We Gon Be Alright.'"

31 Horton-Stallings, *Dirty South Manifesto*, 5.

32 Horton-Stallings, *Dirty South Manifesto*, 5.

33 C. J. Cohen, "Deviance as Resistance," 33.

34 C. J. Cohen, "Deviance as Resistance," 38.

35 Jacobs, *Incidents in the Life of a Slave Girl*, 173–78.

36 Ellison, "Black Femme Praxis," 8.

37 Nash, *Black Body in Ecstasy*, 1.

38 Nash, *Black Body in Ecstasy*, 1.

39 Higginbotham, *Righteous Discontent*, 186; Hammonds, "Black (W)holes and the Geometry of Black Female Sexuality," 175.

40 Spillers, "Mama's Baby, Papa's Maybe," 65.

41 Liboiron and Lepawsky, *Discard Studies*, 143.

42 Virginia Department of Environmental Quality, *Annual Solid Waste Report for CY2022*.

43 W. J. Wright, "As Above, So Below," 794.

44 Charlotte County Sheriff's Office, "Waste Management's Waste Watch Program."

45 McMillan, *Embodied Avatars*, 7.

46 Roane, "Plotting the Black Commons"; Hosbey and Roane, "Totally Different Form of Living."

47 These straight lines of heterosexual sight can't help but produce the white biocentric bodies of property ("man" and "woman"). Uncivilized and unpropertied, the animalization of Blackness is thus the experimental grounds of "biology." Too "animal" to be civilizationally gendered, Black people's genders are made further incoherent by "straight" sight. More like Black holes in un/gendered thought, "Black women" are turned into *voids*—epistemically invisible and conceptually dense—only

detected through their effects on a white binary system. See Hammonds, "Black (W)holes," 126–130; Jackson, "'Theorizing in a Void.'"

48 McKittrick, *Demonic Grounds*, xxv.
49 Jackson, "'Theorizing in a Void'"; da Silva, "1 (life) ÷ 0 (blackness) = ∞ − ∞ or ∞ / ∞"; Hammonds, "Black (W)holes."
50 See Armiero, *Wasteocene*, 12–13, 47–49; Roane, "Plotting the Black Commons."
51 See Shabazz, "'We Gon Be Alright.'"
52 Snorton, "Referential Sights and Slights," 176.
53 Stallings, "Hip Hop and the Black Ratchet Imagination," 136.
54 Alston, "Why Virginia Is Responsible for Hip-Hop's Fixation with Originality."
55 Snorton, *Black on Both Sides*, 83.
56 Haley, *No Mercy Here*.
57 DeBenedetto, "Bed-Stuy Streets Some of the City's Dirtiest."
58 Schuller, *Biopolitics of Feeling*.
59 Hartman, *Scenes of Subjection*, 4.
60 Hartman, *Scenes of Subjection*, 4.
61 Ellison, "Black Femme Praxis," 8.
62 Mutu, *Wangechi Mutu: A Fantastic Journey*; Monteiro, *The Prophecy*; Petry, *The Street*; R. Wright, *Native Son*.
63 Jordan, *Soldier: A Poet's Childhood*, 75.
64 Petry, *The Street*, 1–2.
65 Iheka, *African Ecomedia*, 62.
66 Heynen and Ybarra, "On Abolition Ecologies," 28.
67 Here I'm drawing on the early work of Frank Wilderson and the way Tiffany Lethabo King puts it to use. Wilderson writes about the way the clearing is not only a noun but a verb that labored violently across the landscape and the body of the "other." See Wilderson, *Red, White and Black*, 207. Putting that into the context of settler-colonial mapping, Tiffany Lethabo King wrenches open a space of violent articulation in which Indigenous genocide and enslavement share the violent force of white *conquest*. See King, *Black Shoals*. If, as waste historian Susan Strasser argues in *Waste and Want*, cleaning is an epistemological concern with order, then cleaning is actually a force that disavows its violence, a force that simultaneously inoculates the settler and his grammars to the violence of his own epistemologies. Browne's "Black gaze" repels the force of cleaning (*Dark Matters*, 58–62).
68 Campt, *Black Gaze*, 36.
69 Jackson, *Becoming Human*, 60.
70 Jackson, *Becoming Human*, 61.
71 R. Wright, *Native Son*, 12.
72 R. Wright, *Native Son*, ix.
73 R. Wright, *Native Son*, ii.

74 Critically, Jackson notes, "Owning property is an emblem of white patri-archal masculinity," a genre of man to which Paul D could never belong. See Jackson, *Becoming Human*, 61.

75 Jackson, *Becoming Human*, 214.

76 Though as of 2015, the regulation specifies *motor* vehicles. See New York City Department of Sanitation, *Summary of Sanitation Rules and Regulations*, 54.

77 Iheka, *African Ecomedia*, 57–56.

78 Here, Cajetan Iheka is siting Stephanie Newell, *Histories of Dirt*.

79 Solomon and Wool, "Waste Is Not a Metaphor for Racist Dispossession."

80 Purnell, *Fighting Jim Crow in the County of Kings*.

81 "Brooklyn Group Flaunts Debris," *New York Times*, September 16, 1962.

82 "Brooklyn Group Flaunts Debris."

83 "Brooklyn Group Flaunts Debris."

84 "Brooklyn Group Flaunts Debris." See King, *Black Shoals*, 26–73.

85 Woodsworth, *Battle for Bed Stuy*; C. Taylor, *Fight the Power*; Wilder, *Covenant with Color*; Connolly, *Ghetto Grows in Brooklyn*.

86 Wilder, *Covenant with Color*.

87 Purnell, *Fighting Jim Crow in the County of Kings*.

88 Moynihan, *Negro Family*.

89 McKittrick, *Demonic Grounds*, xiv.

90 McMillan, *Embodied Avatars*, 8.

91 See Hartman, "Belly of the World."

92 McKittrick, *Demonic Grounds*, xvii.

93 Roberts, *Killing the Black Body*, 21.

94 Roberts, *Killing the Black Body*, 21.

95 Chen, *Animacies*, 26.

96 Nelson et al., "Mapping Inequality."

97 Purnell, "Taxation Without Sanitation Is Tyranny," 58.

98 Purnell, "Taxation Without Sanitation Is Tyranny," 53–54.

99 Purnell, "Taxation Without Sanitation Is Tyranny," 67.

100 Purnell, *Fighting Jim Crow in the County of Kings*, 149.

101 Harking back to chapter 2, we can think of this as a form of "capture." Following James Scott's *Seeing Like a State*, the administrative gaze of the state, such as the redlining maps of Bed-Stuy, justify racism by making Black residents' actions, forms of value, and valuation illegible. Similarly here, the administrative gaze of the DSNY obscures residents' claims, ensuring that residents' complaints about DSNY practices are not seen as credible.

102 Purnell, "Taxation Without Sanitation Is Tyranny," 71.

103 Purnell, "Taxation Without Sanitation Is Tyranny"; Purnell, *Fighting Jim Crow in the County of Kings*.

104 There is a long sociological tradition that writes about Black communi-ties by attending to "the street." Du Bois wrote about the "general

movement from the alleys to the streets and from the back to the front streets" as being notable in the study of Black life in Philadelphia's Seventh Ward. See Du Bois, *Philadelphia Negro*, 306. For white sociologists, however, the concern with how Black people move and act on the streets of cities would come to take on new meanings, turning culture into the problem of Black poverty, *not* the lack of economic opportunities. See Lewis, "The Culture of Poverty." Though declared to have had a "resurgence" by the *New York Times* in the twenty-first century (see P. Cohen, "'Culture of Poverty' Makes a Comeback"), the culture of poverty thesis has long had a persistent hold over the imagination of poor life on the street despite its many critics. (On critiques of the culture of poverty, see Ryan, *Blaming the Victim*; Katz, *Undeserving Poor*; O'Connor, *Poverty Knowledge*; Royster, *Race and the Invisible Hand*; hooks, *We Real Cool*.) These accounts of street life take the production of the spatial unit of "the street" for granted, not only deprioritizing class as a racial-spatial relation (see Rothstein, *Color of Law*) but continuously disappearing women (see McKittrick, *Demonic Grounds*). I see this trend in urban anthropology and sociology as situated within the intersections of environmental determinist arguments about place and behavior, the culture of poverty, and the fetishizing of the war on drugs, culminating in a racializing project that turns Black sociality into poverty porn in the social sciences. For more on these racist politics that frame poor Black environments as "blight" and criminalize antisocial behavior on the street, see Gregory, "Race, Rubbish, and Resistance"; Gregory, *Black Corona*; Kelley, "'We Are Not What We Seem'"; O'Connor, *Poverty Knowledge*.

105 M. Alexander, *New Jim Crow*; Gilmore, *Golden Gulag*; Roberts, *Killing the Black Body*.

106 Purnell, "Taxation Without Sanitation Is Tyranny," 73.

107 Purnell, "Taxation Without Sanitation Is Tyranny," 73.

108 McKittrick, *Dear Science and Other Stories*; Harney and Moten, *Undercommons*.

109 McKittrick, *Dear Science and Other Stories*.

110 Purnell, *Fighting Jim Crow in the County of Kings*, 152, 154.

111 Purnell, *Fighting Jim Crow in the County of Kings*, 152.

112 C. J. Cohen, "Deviance as Resistance," 30.

113 Allewaert, *Ariel's Ecology*, 49.

DISPOSAL

1 Pellow, *Garbage Wars*, 1.
2 EPA, *Municipal Solid Waste Generation*.
3 Pellow, *Garbage Wars*, 1.

4   Melosi, *Garbage in the Cities*; Strasser, *Waste and Want*; Pellow, *Garbage Wars*; Zimring, *Clean and White*; Strach and Sullivan, *Politics of Trash*; Rathje and Murphy, *Rubbish!*

5   Gidwani and Maringanti, "Waste-Value Dialectic," 118.

6   Laporte, *History of Shit*, 56.

7   Rathje and Murphy, *Rubbish!*, 12.

8   *Toxic Wastes and Race in the United States*, 4.

9   *Toxic Wastes and Race in the United States*, 24.

10   As Elizabeth Bernstein and Janet Jakobsen remind us in the introduction to their edited volume *Paradoxes of Neoliberalism*, neoliberalism is full of death-dealing paradoxes: during the COVID-19 pandemic, "the common designation of some forms of labor as 'essential,' whereas the workers in question (disproportionately Black and brown, migrants, and female) were treated as disposable" (1).

11   Mills, "Black Trash," 80.

12   See Tuck and Yang, "Decolonization Is Not a Metaphor"; Hartman, *Scenes of Subjection*.

13   Mills, "Black Trash," 84. For more on the way "white trash" marks a history of race and class relations, see Isenberg, *White Trash*.

14   Bullard, *Dumping in Dixie*, 97.

15   Bullard, *Dumping in Dixie*, 98.

CHAPTER FOUR. BLACK REFRACTIONS

1   Hartman, "Venus in Two Acts," 11.

2   Hartman, "Venus in Two Acts," 11.

3   Kelley, "'We Are Not What We Seem'"; Stamatopoulou-Robbins, *Waste Siege*; Millar, *Reclaiming the Discarded*.

4   See Fanon, *Wretched of the Earth*; Robinson, *Black Marxism*; Woods, *Development Arrested*; Gilmore, *Golden Gulag*.

5   Millar, *Reclaiming the Discarded*; Simone, "People as Infrastructure"; Simone, *Improvised Lives*; Von Schnitzler, *Democracy's Infrastructure*.

6   N. Eisenberg-Guyot, "On How to Live While Being Thrown Away," 184. See Nixon, *Slow Violence*.

7   N. Eisenberg-Guyot, "On How to Live While Being Thrown Away," 184–86.

8   Department of City Planning, "Bedford-Stuyvesant South Rezoning Proposal"; Department of City Planning, "Bedford-Stuyvesant North Rezoning Proposal."

9   DeBenedetto, "Bed-Stuy Streets Some of the City's Dirtiest."

10   Solomon, "Ghetto Is a Gold Mine."

11   Jordan, *Living Room*.

12 Solomon, "Ghetto Is a Gold Mine."

13 Bauman, *Wasted Lives*.

14 Lutz and Collins, "Photograph as an Intersection of Gazes."

15 A relation of labor (see Federici, *Caliban and the Witch*), a site of racial domination (see A. Y. Davis, *Women, Race and Class*; Lorde, *Sister Outsider*), a central pillar in the matrix of domination (see Collins, *Black Feminist Thought*; Crenshaw, "Mapping the Margins"), a practice (see Butler, *Gender Trouble*), a biocentric marker of colonial dispossession (see Wynter, "From 'Beyond Miranda's Meanings'"; Wynter, "Unsettling the Coloniality of Being/Power/Truth/Freedom"), and a colonial invention (see Oyěwùmí, *Invention of Women*). Thank you so much to Nadja Eisenberg-Guyot for this formulation and for our conversations about Saidiya Hartman's work and the way ethnographic writing is a challenge of violent proportions.

16 I'm thinking of animation here the way Mel Chen uses animacies to think through the gendered and sexualized discourses that permeate animation of dangerous intimacies—the toxic, the parasitic, and so on (see Chen, *Animacies*).

17 If we are to be intersectional feminists in our attention to matter as well, we must be careful how we deploy Western tropes of matter. In particular, the way description slides into empiricism is a *matter of empire*, and regimes of racial signification risk deploying descriptions that stabilize "the savage slot" (see Trouillot, *Silencing the Past*) as a carceral geography in our language (see Schept, "(Un)seeing Like a Prison").

18 In Angela Davis's *Women, Race and Class*, she describes Black women's role in the community of slaves to unsettle the reading of these scenes as heteropatriarchal. Davis's point is that African modalities of family care are subsumed under white patriarchal protocols and significations but were also/actually something else, something rendered hard to see by the domestication of Black genders to white patriarchy.

19 Spillers, "Mama's Baby, Papa's Maybe," 67.

20 Snorton, *Black on Both Sides*; Bey, *Cistem Failure*.

21 For more on the structural production of the captive maternal, see James, "Captive Maternal and Abolitionism"; Williamson, *Scandalize My Name*.

22 Hosbey and Roane, "Totally Different Form of Living."

23 Ferguson, *Aberrations in Black*.

24 McKittrick, *Demonic Grounds*; Hawthorne, "Black Matters Are Spatial Matters."

25 Hicks, "Mayor Announces New Assault on Graffiti."

26 A corollary of this argument is implicit in the pro-gentrification stance: that it is the new (white, upper-middle-class) residents of a gentrifying neighborhood that can "restore" the neighborhood to its "true" value by

preserving its history—these new residents are the proper stewards of value. Again, the assumption is that Black people don't—can't—care for property and, in their lack of care, cause property to decay.

27 Smith, "Giuliani Time," 1.

28 Solomon, "Ghetto Is a Gold Mine."

29 Campt, *Listening to Images*.

30 Quashie, *Sovereignty of Quiet*, 8.

31 Patterson, *Slavery and Social Death*, 38.

32 Kaplan, "Love and Violence/Maternity and Death," 112.

33 C. J. Cohen, "Deviance as Resistance," 29.

34 C. J. Cohen, "Deviance as Resistance," 42.

35 See Walcott, *On Property*. Bios-mythos is a term fashioned by Sylvia Wynter to describe the way that biology becomes a myth of the Eurocentric genre of the "human." For more, see Wynter, "Unsettling the Coloniality of Being/Power/Truth/Freedom"; McKittrick, *Sylvia Wynter*, 1–8.

36 Morgensen, "Biopolitics of Settler Colonialism"; Locke, *Second Treatise of Government*.

37 Walcott, *On Property*.

38 See J. T. Roane's work on queer urbanism, *Dark Agoras*, 147–80.

39 Keeling, *Queer Times, Black Futures*, xiii.

40 Keeling, *Queer Times, Black Futures*, xiii.

41 This particular formation draws inspiration from Kassidy Jones. At the Black Environmentalism Symposium in October 2022 at Yale University, Kassidy spoke about what she calls "literary tending." While Sal is not tending to literature, I see his generative attention to things as a kind of tending that, like Kassidy explains, allows for the proliferation of reading, not an overdetermined one.

JUNK

1 The "oppositional gaze" was a term coined by the late bell hooks in her book *Black Looks: Race and Representation*. Here I'm also taking inspiration from the way Willie Jamaal Wright uses music to "see" Black environmental politics (see Wright, "As Above, So Below"). I'm also drawing on Simone Browne's resituating of the panopticon's "gaze from nowhere" to a gaze made on a slave ship (see Browne, *Dark Matters*) and Tina Campt's meditation on how Black artists change the way we *see* (see Campt, *Black Gaze*).

2 J. Byrd, "To Hear the Call and Respond," 337.

3 Vasudevan, "Intimate Inventory of Race and Waste"; Sharpe, *In the Wake*; K. Simmons, "Settler Atmospherics"; Fanon, *Wretched of the Earth*; Op-

perman, "Need for a Black Feminist Climate Justice"; Roane, *Dark Agoras*; Hosbey and Roane, "Totally Different Form of Living"; W. J. Wright, "The Morphology of Marronage"; W. J. Wright, "As Above, So Below"; Diouf, *Slavery's Exiles*; Kelley, *Freedom Dreams*.

4   Woods, *Development Arrested*, 20.

5   Woods, *Development Arrested*, 20.

6   Shabazz, "'We Gon Be Alright,'" 451. "Furious styles," here, marks a reference to Jeff Chang's history of the hip-hop generation's movement-defining aesthetics. See Chang, *Can't Stop Won't Stop*.

7   Shabazz, "'We Gon Be Alright,'" 452.

8   Silkenat, *Scars on the Land*.

9   King, *Black Shoals*.

10   J. G. Williams, *Crossing Bar Lines*, 6.

CONCLUSION

1   Pristin, "In Bedford-Stuyvesant, the Boom Remains a Bust."

2   In Katherine McKittrick's *Dear Science and Other Stories*, she argues, "Metaphors offer an (entwined material and imagined) future that has not arrived and the future we live and have already lived through" (11).

3   Here I'm referring to improvisation as a tradition of both music and living. In the final chapter of Robin Kelley's *Freedom Dreams*, he talks about the political imaginations of Black artists and surrealists as something born of the violence wrought on the Black diaspora: "We are, after all, talking about cultures that valued imagination, improvisation, and verbal agility, from storytelling, preaching and singing to toasting and the dozens" (205). Extending Kelley's analysis, I see this as central to the everyday practice of Black regard while being discarded. In "To Extend," AbdouMaliq Simone also builds on this argument. Thinking through urban improvisation as a modality produced within survival, he says, "At the same time, they engage their surrounds as a field of improvisations, of sites and functions that could be composed in different ways" (1136). There are so many other scholars and practitioners to cite here I can only modestly signal to them: Nyong'o, *Afro-Fabulations*; McMillan, *Embodied Avatars*; Hartman, *Lose Your Mother*. See also the science fiction of Octavia Butler, *Kindred*, *Parable of the Sower*, *Parable of the Talents*, *Wild Seed*, *Dawn*, and *Fledgling*; and N. K. Jemisin's *The Fifth Season*, *The Obelisk Gate*, and *The Stone Sky*. See also the artwork of Wangechi Mutu and Fabrice Monteiro; and listen: Ornette Coleman, *Something Else!!!!* and *Tomorrow Is the Question!*; Thelonious Monk, *It's Monk's Time* and *Solo Monk*; Miles Davis, *Blue Period*; and Ma Rainey, "See, See Rider Blues."

4   See Iheka, *African Ecomedia*; Woods, *Development Arrested*.

5   Keeling, *Queer Times, Black Futures*.

6   Jackson, *Becoming Human*.

7   Da Silva, "Toward a Black Feminist Poethics," 85.

8   Yusoff, *Billion Black Anthropocenes or None*; Jackson, *Becoming Human*.

9   Da Silva, "Toward a Black Feminist Poethics," 82.

10 See daSilva, "Toward a Black Feminist Poethics."

11 Da Silva, "Toward a Black Feminist Poethics," 84.

12 Da Silva, "Toward a Black Feminist Poethics," 84.

13 Here I'm thinking about the way a white audience could not receive the work of Zora Neale Hurston. Her refusal to either misportray formerly enslaved Black Southerners' speech (see Hurston, *Mules and Men*) or simply see ethnography as the only way to study Black life lead her to different genres of nonfiction. Most notably she authored *Barracoon* based on her interviews with Oluale Kossola in 1927, the last known survivor of the Middle Passage. See also Hurston et.al, *Their Eyes Were Watching God*.

14 Jacobs-Huey, "The Natives Are Gazing and Talking Back"; Morris, introduction to *Can the Subaltern Speak?*; Spivak, "Can the Subaltern Speak"; Tuck and Yang, "R-Words." Thank you to Katherine Detwiler for this formation of anthropology as mining. Sitting on her dissertation committee taught me a lot about the importance of thinking about metaphor as more than metaphor in the field. See Detwiler, "Science Fell in Love with the Chilean Sky."

15 Visweswaran, *Fictions of Feminist Ethnography*, 10.

16 Visweswaran, *Fictions of Feminist Ethnography*, 13.

17 Jobson, "Case for Letting Anthropology Burn," 261; Nyong'o, *Afro-Fabulations*, 26.

18 Da Silva, "Toward a Black Feminist Poethics," 81.

19 Lorde, "Poetry Is Not a Luxury."

20 Shange, *Progressive Dystopia*, 9.

21 Butler, *Kindred*, 9.

22 Dana, recounting the trauma of losing her arm after her last trip through the wall in the antebellum South in *Kindred* (264). McKittrick cites this moment, too, as a way to begin the (un)geographic subject position that Black women bear inside plantation futures. See McKittrick, *Demonic Grounds*.

23 This is a reference to so many things, but specifically to my uncle's friend's cousin Dan, who's the most dapper man I know and who runs a record shop in Bed-Stuy that has somehow been there for thirty years and is still kicking.

24 Walker, "The Notorious B.I.G. Mural."

25 The Notorious B.I.G., *Ready to Die*.

26 Jackson, *Becoming Human*, 64–69.

27 Sharpe, *Monstrous Intimacies*, 2; Hartman, *Scenes of Subjection*, 3–4.

28   Fanon, *Black Skin, White Masks*, 82.

29   See da Silva, "Toward a Black Feminist Poethics"; Snorton, *Black on Both Sides*.

30   Da Silva, "Toward a Black Feminist Poethics," 87.

31   Da Silva, "Toward a Black Feminist Poethics," 86.

32   Davis et al., *Abolition. Feminism. Now.*.

33   Da Silva, "Toward a Black Feminist Poethics," 84.

34   Da Silva, "Toward a Black Feminist Poethics," 92–93.

Bibliography

Adams, Terri M., and Douglas B. Fuller. "The Words Have Changed but the Ideology Remains the Same: Misogynistic Lyrics in Rap Music." *Journal of Black Studies* 36, no. 6 (2006): 938–57.

Agard-Jones, Vanessa. "Bodies in the System." *Small Axe: A Caribbean Journal of Criticism* 17, no. 3 (2013): 182–92. https://doi.org/10.1215/07990537-2378991.

Agard-Jones, Vanessa. "What the Sands Remember." GLQ: *A Journal of Lesbian and Gay Studies* 18, no. 2–3 (2012): 325–46. https://doi.org/10.1215/10642684-1472917.

Ahuja, Neel. "Intimate Atmospheres." GLQ: *A Journal of Lesbian and Gay Studies* 21, no. 2–3 (2015): 365–85. https://doi.org/10.1215/10642684-2843227.

Alexander, Catherine, and Joshua Reno, eds. *Economies of Recycling: The Global Transformation of Materials, Values and Social Relations.* London: Zed Books, 2012.

Alexander, M. Jacqui. "Redrafting Morality: The Postcolonial State and the Sexual Offences Bill of Trinidad and Tobago." In *Third World Women and the Politics of Feminism*, edited by Chandra Talpade Mohanty, Ann Russo, and Torres Lourdes. Bloomington: Indiana University Press, 1991.

Alexander, Michelle. *The New Jim Crow: Mass Incarceration in the Age of Colorblindness.* New York: New Press, 2020.

Allen, Jafari S. "Black/Queer/Diaspora at the Current Conjuncture." GLQ: *A Journal of Lesbian and Gay Studies* 18, no. 2–3 (2012): 211–48. https://doi.org/10.1215/10642684-1472872.

Allen, Michael Thad, and Gabrielle Hecht, eds. "Technology, Politics, and National Identity in France." In *Technologies of Power: Essays in Honor of Thomas Parke Hughes and Agatha Chipley Hughes.* Cambridge, MA: MIT Press, 2001. https://doi.org/10.7551/mitpress/6679.003.0010.

Allen, Ryan W., Michael H. Criqui, Ana V. Diez Roux, et al. "Fine Particulate Matter Air Pollution, Proximity to Traffic, and Aortic Atherosclerosis." *Epidemiology* 20, no. 2 (2009): 254–64. https://doi.org/10.1097/EDE.0b013e31819644cc.

Allewaert, Monique. *Ariel's Ecology: Plantations, Personhood, and Colonialism in the American Tropics.* Minneapolis: University of Minnesota Press, 2013.

Alston, Trey. "Why Virginia Is Responsible for Hip-Hop's Fixation with Originality." *Revolt* (blog). May 24, 2018. https://www.revolt.tv/article/2018-05-24/96069/why-virginia-is-responsible-for-hip-hops-fixation-with-originality/.

Appadurai, Arjun. *The Social Life of Things: Commodities in Cultural Perspective.* 7th ed. Cambridge: Cambridge University Press, 2009.

Armiero, Marco. "The Case for the Wasteocene." *Environmental History* 26, no. 3 (2021): 425–30. https://doi.org/10.1093/envhis/emab014.003.

Armiero, Marco. *Wasteocene: Stories from the Global Dump.* Cambridge Elements in Environmental Humanities. Cambridge: Cambridge University Press, 2021.

Armiero, Marco, and Massimo De Angelis. "Anthropocene: Victims, Narrators, and Revolutionaries." *South Atlantic Quarterly* 116, no. 2 (2017): 345–62. https://doi.org/10.1215/00382876-3829445.

Barad, Karen. "Posthumanist Performativity: Toward an Understanding of How Matter Comes to Matter." *Signs: Journal of Women in Culture and Society* 28, no. 3 (2003): 801–31. https://doi.org/10.1086/345321.

Barker, Joanne. "Territory as Analytic." *Social Text* 36, no. 2 (2018): 19–39. https://doi.org/10.1215/01642472-4362337.

Barry, Andrew. *Political Machines: Governing a Technological Society.* London: Athlone Press, 2001.

Barry, Andrew. "Technological Zones." *European Journal of Social Theory* 9, no. 2 (2006): 239–53. https://doi.org/10.1177/1368431006063343.

Bataille, Georges. *Visions of Excess: Selected Writings, 1927–1939.* Edited by Allan Stoekl, translated by Allan Stoekl, with Carl R. Lovitt and Donald M. Leslie Jr. Minneapolis: University of Minnesota Press, 1985.

Bauman, Zygmunt. *Wasted Lives: Modernity and Its Outcasts.* 2004. Reprint, Cambridge: Polity Press, 2011.

Beal, Frances. "Double Jeopardy: To Be Black and Female." In *The Black Woman: An Anthology,* edited by Toni Cade Bambara. New York: Washington Square Press, 1970.

Benjamin, Ruha. *Race After Technology: Abolitionist Tools for the New Jim Code.* Cambridge: Polity Press, 2020.

Benjamin, Walter. "Theses on the Philosophy of History." In *Illuminations: Essays and Reflections,* edited by Hannah Arendt, translated by Harry Zohn. New York: Schocken Books, 2012.

Bennett, Jane. *Vibrant Matter: A Political Ecology of Things.* Durham, NC: Duke University Press, 2010.

Bennett, Joshua. *Being Property Once Myself: Blackness and the End of Man.* Cambridge, MA: Harvard University Press, 2022.

Bernstein, Elizabeth. "Militarized Humanitarianism Meets Carceral Feminism: The Politics of Sex, Rights, and Freedom in Contemporary Antitrafficking Campaigns." *Signs: Journal of Women in Culture and Society* 36, no. 1 (2010): 45–71. https://doi.org/10.1086/652918.

Bernstein, Elizabeth, and Janet R. Jakobsen, eds. *Paradoxes of Neoliberalism: Sex, Gender and Possibilities for Justice.* London: Routledge, 2022.

Berthold, Dana. "Tidy Whiteness: A Genealogy of Race, Purity, and Hygiene." *Ethics and the Environment* 15, no. 1 (2010): 1. https://doi.org/10.2979/ete.2010.15.1.1.

Bey, Marquis. *Cistem Failure: Essays on Blackness and Cisgender.* Durham, NC: Duke University Press, 2022.

Biehl, João, and Torben Eskerod. *Vita: Life in a Zone of Social Abandonment.* Berkeley: University of California Press, 2013.

Bledsoe, Adam. "Marronage as a Past and Present Geography in the Americas." *Southeastern Geographer* 57, no. 1 (2017): 30–50. https://doi.org/10.1353/sgo.2017.0004.

Bonhomme, Edna. "Contagion on the Plantation." *E-flux* Architecture, May 2022. https://www.e-flux.com/architecture/sick-architecture/453874/contagion-on-the-plantation/.

Bowker, Geoffrey C., and Susan Leigh Star. *Sorting Things Out: Classification and Its Consequences.* Cambridge, MA: MIT Press, 2000. https://doi.org/10.7551/mitpress/6352.001.0001.

Braidotti, Rosi. *Metamorphoses: Towards a Materialist Theory of Becoming.* Cambridge: Polity Press, 2002.

Brand, Dionne. *Inventory.* Toronto: McClelland and Stewart, 2006.

Brenner, Neil, Jamie Peck, and Nik Theodore. "Variegated Neoliberalization: Geographies, Modalities, Pathways." *Global Networks* 10, no. 2 (2010): 182–222. https://doi.org/10.1111/j.1471-0374.2009.00277.x.

Brenner, Neil, and Nik Theodore. "Neoliberalism and the Urban Condition." *City* 9, no. 1 (2005): 101–7. https://doi.org/10.1080/13604810500092106.

Brown, Bill. "Thing Theory." *Critical Inquiry* 28, no. 1 (2001): 1–22.

Brown, Jayna. "Hip Hop, Pleasure, and Its Fulfillment." *Palimpsest: A Journal on Women, Gender, and the Black International* 2, no. 2 (2013): 147–50. https://doi.org/10.1353/pal.2013.0015.

Browne, Simone. *Dark Matters: On the Surveillance of Blackness.* Durham, NC: Duke University Press, 2015.

Bruce, La Marr Jurelle. *How to Go Mad Without Losing Your Mind: Madness and Black Radical Creativity.* Durham, NC: Duke University Press, 2021.

Bullard, Robert D. *Dumping in Dixie: Race, Class and Environmental Quality.* 3rd ed. London: Routledge, 2019.

Bullard, Robert D. "Environmental Blackmail in Minority Communities." In *Race and the Incidence of Environmental Hazards: A Time for Discourse*, edited by Bunyan I. Bryant and Paul Mohai. Abingdon: Routledge, 2019.

Bullard, Robert D. "Environmental Justice in the 21st Century: Race Still Matters." *Phylon (1960–)* 49, no. 3/4 (2001): 151–71. https://doi.org/10.2307/3132626.

Burden-Stelly, Charisse. "Modern U.S. Racial Capitalism." *Monthly Review*, July 2020, 8–20. https://doi.org/10.14452/MR-072-03-2020-07_2.

Butler, Judith. *Gender Trouble: Feminism and the Subversion of Identity.* New York: Routledge, 2006.

Butler, Octavia E. *Bloodchild and Other Stories.* 2nd ed. New York: Seven Stories Press, 2005.

Butler, Octavia E. *Dawn.* New York: Grand Central, 2021.

Butler, Octavia E. *Fledgling*. New York: Grand Central, 2022.

Butler, Octavia E. *Kindred*. Garden City, NY: Doubleday, 1979.

Butler, Octavia E. *Parable of the Sower*. New York: Grand Central, 2019.

Butler, Octavia E. *Parable of the Talents*. New York: Grand Central, 2019.

Butler, Octavia E. *Seed to Harvest (Wild Seed, Mind of My Mind, Clay's Ark, Patternmaster)*. New York: Warner Books, 2007.

Butt, Waqas H. *Life Beyond Waste: Work and Infrastructure in Urban Pakistan*. Stanford, CA: Stanford University Press, 2023.

Byrd, Jodi. "To Hear the Call and Respond: Grounded Relationalities and the Spaces of Emergence." *American Quarterly* 71, no. 2 (2019): 337–42.

Byrd, William. "The History of the Dividing Line Betwixt Virginia and North Carolina." In *The Westover Manuscripts*, edited by Edmund Ruffin and Julian Ruffin. Petersburg, VA, 1841.

Camboulives, J., F. Sicardi, and J. Bimar. "Appearance of Coagulation Disorders in the Non-Healthy Newborn Infant [in French]." *Annales de L'anesthesiologie Française* 16, spec. no. 1 (1975): 181–94.

Campt, Tina M. *A Black Gaze: Artists Changing How We See*. Cambridge, MA: MIT Press, 2021.

Campt, Tina M. *Listening to Images*. Durham, NC: Duke University Press, 2017.

Carby, Hazel V. *Imperial Intimacies: A Tale of Two Islands*. London: Verso, 2021.

Caverly, Nicholas L. "Carceral Structures: Financialized Displacement and Captivity in Detroit." *Anthropological Quarterly* 95, no. 2 (2022): 333–61. https://doi.org/10.1353/anq.2022.0018.

Centers for Disease Control and Prevention. "Recommended Actions Based on Blood Lead Level." Last updated August 19, 2024. https://www.cdc.gov/lead-prevention/hcp/clinical-guidance/index.html.

Cervenak, Sarah Jane. *Black Gathering: Art, Ecology, Ungiven Life*. Black Outdoors. Durham, NC: Duke University Press, 2021.

Cervenak, Sarah Jane. *Wandering: Philosophical Performances of Racial and Sexual Freedom*. Durham, NC: Duke University Press, 2014. https://doi.org/10.1215/9780822376347.

Césaire, Aimé. *Discourse on Colonialism*. With a new introduction by Robin D. G. Kelley. New York: Monthly Review Press, 2000.

Chahim, Dean. "The Logistics of Waste: Engineering, Capital Accumulation, and the Growth of Mexico City." *Antipode*, July 2022. https://doi.org/10.1111/anti.12864.

Chalfin, Brenda. "Public Things, Excremental Politics, and the Infrastructure of Bare Life in Ghana's City of Tema." *American Ethnologist* 41, no. 1 (2014): 92–109. https://doi.org/10.1111/amet.12062.

Chang, Jeff. *Can't Stop Won't Stop: A History of the Hip-Hop Generation*. New York: Picador, 2006.

Charlotte County Sheriff's Office. "Waste Management's Waste Watch Program; Reporting Suspicious Activity." *Charlotte County Sheriff's Office* (blog), December 26, 2018. https://ccsoblog.org/2018/12/26/waste-watch/.

Checker, Melissa. *Polluted Promises: Environmental Racism and the Search for Justice in a Southern Town*. New York: New York University Press, 2005.

Checker, Melissa. *The Sustainability Myth: Environmental Gentrification and the Politics of Justice*. New York: New York University Press, 2020.

Chen, Mel Y. *Animacies: Biopolitics, Racial Mattering, and Queer Affect*. Durham, NC: Duke University Press, 2012.

Chowdhry, Geeta, and Mark Beeman. "Situating Colonialism, Race, and Punishment." In *Race, Gender, and Punishment: From Colonialism to the War on Terror*, edited by Mary Bosworth and Jeanne Flavin. New Brunswick, NJ: Rutgers University Press, 2007.

Circular Claims Fall Flat Again: 2022 Update. Washington, DC: Greenpeace, 2022. https://www.greenpeace.org/usa/wp-content/uploads/2022/10/GPUS _FinalReport_2022.pdf.

Clipse. "Virginia." Track 3 on *Lord Willin'*. Produced by the Neptunes. Arista. Released 2002.

Cohen, Cathy J. "Deviance as Resistance: A New Research Agenda for the Study of Black Politics." *Du Bois Review: Social Science Research on Race* 1, no. 1 (2004): 27–45. https://doi.org/10.1017/S1742058X04040044.

Cohen, Cathy J. "Punks, Bulldaggers, and Welfare Queens: The Radical Potential of Queer Politics?" *GLQ: A Journal of Lesbian and Gay Studies* 3, no. 4 (1997): 437–65. https://doi.org/10.1215/10642684-3-4-437.

Cohen, Patricia. "'Culture of Poverty' Makes a Comeback." *New York Times*, October 17, 2010. https://www.nytimes.com/2010/10/18/us/18poverty.html.

Coleman, Ornette. *Something Else!!!! The Music of Ornette Coleman*. Produced by Lester Koenig. Contemporary S7551. Released 1958.

Coleman, Ornette. *Tomorrow Is the Question!* Produced by Lester Koenig. Contemporary M3569. Released 1959.

Collins, Patricia Hill. *Black Feminist Thought: Knowledge, Consciousness, and the Politics of Empowerment*. 2nd ed. New York: Routledge, 2009.

Collins, Patricia Hill. *Black Sexual Politics: African Americans, Gender, and the New Racism*. New York: Routledge, 2004.

Collins, Patricia Hill. "Gender, Black Feminism, and Black Political Economy." *Annals of the American Academy of Political and Social Science* 568 (March 2000): 41–53.

Collins, Patricia Hill. *Intersectionality as Critical Social Theory*. Durham, NC: Duke University Press, 2019.

Collins, Patricia Hill. *Lethal Intersections*. Cambridge: Polity Press, 2024.

Connolly, Harold X. *A Ghetto Grows in Brooklyn*. New York: New York University Press, 1977.

Coren, Michael. "50 Years of Research Shows There Is No Safe Level of Childhood Lead Exposure." *Pulitzer Center*, June 16, 2022. https://pulitzercenter.org /stories/50-years-research-shows-there-no-safe-level-childhood-lead -exposure.

Cowen, Deborah. *The Deadly Life of Logistics: Mapping Violence in Global Trade.* Minneapolis: University of Minnesota Press, 2014.

Cox, Jeremy. "Naval Power Plant Proposal Tests Virginia on Environmental Justice." *Bay Journal*, January 5, 2021. https://www.bayjournal.com/news/energy/naval-power-plant-proposal-tests-virginia-on-environmental-justice/article_8b8db992-3a5a-11eb-9a48-d78ac19efa91.html.

Crenshaw, Kimberlé Williams. "Beyond Racism and Misogyny: Black Feminism and 2 Live Crew." In *Words That Wound: Critical Race Theory, Assaultive Speech, and the First Amendment*, edited by Mari J. Matsuda, Charles R. Lawrence III, Richard Delgado, and Kimberlé Williams Crenshaw. New York: Routledge, 2018. https://doi.org/10.4324/9780429502941.

Crenshaw, Kimberle Williams. "Mapping the Margins: Intersectionality, Identity Politics, and Violence Against Women of Color." *Stanford Law Review* 43, no. 6 (1991): 1241. https://doi.org/10.2307/1229039.

da Silva, Denise Ferreira. "1 (life) ÷ 0 (blackness) = ∞ − ∞ or ∞ / ∞: On Matter Beyond the Equation of Value." *E-flux* 79 (February 2017). https://www.e-flux.com/journal/79/94686/1-life-0-blackness-or-on-matter-beyond-the-equation-of-value/.

da Silva, Denise Ferreira. "Toward a Black Feminist Poethics: The Quest(ion) of Blackness Toward the End of the World." *Black Scholar* 44, no. 2 (2014): 81–97. https://doi.org/10.1080/00064246.2014.11413690.

Davies, Thom. "Slow Violence and Toxic Geographies: 'Out of Sight' to Whom?" *Environment and Planning C: Politics and Space* 40, no. 2 (2022): 409–27. https://doi.org/10.1177/2399654419841063.

Davies, Thom. "Toxic Space and Time: Slow Violence, Necropolitics, and Petrochemical Pollution." *Annals of the American Association of Geographers* 108, no. 6 (2018): 1537–53. https://doi.org/10.1080/24694452.2018.1470924.

Davis, Angela Y. *Are Prisons Obsolete?* New York: Seven Stories Press, 2003.

Davis, Angela Y. *Women, Race and Class.* New York: Vintage Books, 1983.

Davis, Angela Y., Gina Dent, Erica R. Meiners, and Beth E. Richie. *Abolition. Feminism. Now.* Chicago: Haymarket Books, 2022.

Davis, Dána-Ain. "The Politics of Reproduction: The Troubling Case of Nadya Suleman and Assisted Reproductive Technology[*]." *Transforming Anthropology* 17, no. 2 (2009): 105–16. https://doi.org/10.1111/j.1548-7466.2009.01061.x.

Davis, Dána-Ain. *Reproductive Injustice: Racism, Pregnancy, and Premature Birth.* New York: New York University Press, 2019.

Davis, Miles. *Blue Period.* Produced by Bob Weinstock. Prestige PRLP 140. Released 1953.

Davis, Miles. *Miles Davis Quartet.* Produced by Bob Weinstock. Prestige PRLP 161. Released 1954.

DeBenedetto, Paul. "Bed-Stuy Streets Some of the City's Dirtiest, Report Says." DNA *Info*, January 13, 2014. https://www.dnainfo.com/new-york/20140113/bed-stuy/bed-stuy-streets-some-of-citys-dirtiest-report-says/.

Denise, DJ Lynnée. "Dancing Between Worlds: An Interview with Saul Williams." *Los Angeles Review of Books*, March 5, 2019. https://lareviewofbooks.org /article/dancing-between-worlds-an-interview-with-saul-williams.

Denise, DJ Lynnée. "Where House Found a Home: The Story of South African House Music." *5 Magazine*, April 2012.

Department of City Planning. "Bedford-Stuyvesant North Rezoning Proposal." Department of City Planning, New York, 2012. https://www.nyc.gov /assets/planning/download/pdf/plans/bedford-stuyvesant-north/bed _stuy_north.pdf.

Department of City Planning. "Bedford-Stuyvesant South Rezoning Proposal." Department of City Planning, New York, 2007. https://www.nyc.gov /assets/planning/download/pdf/plans/bedford-stuyvesant/bed_stuy .pdf.

DeSouza, Priyanka, Danielle Braun, Robbie M. Parks, Joel Schwartz, Francesca Dominici, and Marianthi-Anna Kioumourtzoglou. "Nationwide Study of Short-Term Exposure to Fine Particulate Matter and Cardiovascular Hospitalizations Among Medicaid Enrollees." *Epidemiology* 32, no. 1 (2021): 6–13. https://doi.org/10.1097/EDE.0000000000001265.

Detwiler, Katheryn. "Science Fell in Love with the Chilean Sky." New York: New School for Social Research, 2022.

Dillon, Lindsey, and Julie Sze. "Police Power and Particulate Matters: Environmental Justice and the Spatialities of In/Securities in U.S. Cities." *English Language Notes* 54, no. 2 (2016): 13–23.

Diouf, Sylviane A. *Slavery's Exiles: The Story of the American Maroons.* New York: New York University Press, 2014.

"Dismal Swamp Land Company Articles of Agreement, 3 November 1763." Founders Online. National Archives, Washington, DC. https://founders .archives.gov/documents/Washington/02-07-02-0163.

Douglas, Mary. *Purity and Danger: An Analysis of Concept of Pollution and Taboo.* London: Routledge, 2005.

Douglass, Frederick. *The Heroic Slave: A Cultural and Critical Edition.* Edited by Robert S. Levine, John Stauffer, and John R. McKivigan. New Haven, CT: Yale University Press, 2015.

Du Bois, W. E. B. *Black Reconstruction in America: 1860–1880.* New York: Free Press, 1998.

Du Bois, W. E. B. *The Philadelphia Negro: A Social Study.* Philadelphia: University of Pennsylvania Press, 1996. https://doi.org/10.9783/9780812201802.

"Dumping Grounds." *The Economist*, February 18, 1999. https://www.economist .com/united-states/1999/02/18/dumping-grounds.

Durham, Aisha, Brittney C. Cooper, and Susana M. Morris. "The Stage Hip-Hop Feminism Built: A New Directions Essay." *Signs: Journal of Women in Culture and Society* 38, no. 3 (2013): 721–37. https://doi.org/10.1086 /668843.

Edwards, Paul. "Infrastructure and Modernity: Force, Time, and Social Organization in the History of Sociotechnical Systems." In *Modernity and Technology*, edited by Thomas J. Misa, Philip Brey, and Andrew Feenberg. Cambridge, MA: MIT Press, 2002. https://doi.org/10.7551/mitpress/4729 .001.0001.

Eisenberg-Guyot, Jerzy, Caislin Firth, Marieka Klawitter, and Anjum Hajat. "From Payday Loans to Pawnshops: Fringe Banking, the Unbanked, and Health." *Health Affairs* 37, no. 3 (2018): 429–37. https://doi.org/10.1377 /hlthaff.2017.1219.

Eisenberg-Guyot, Jerzy, Trevor Peckham, Sarah B. Andrea, Vanessa Oddo, Noah Seixas, and Anjum Hajat. "Life-Course Trajectories of Employment Quality and Health in the U.S.: A Multichannel Sequence Analysis." *Social Science and Medicine* 264 (November 2020): 113327. https://doi.org /10.1016/j.socscimed.2020.113327.

Eisenberg-Guyot, Nadja. "On How to Live While Being Thrown Away: Black People Who Use Drugs and the Politics of Anti-Disposability, North Philadelphia, circa 2007 to 2010." *City and Society* 35, no. 3 (2023): 180–90. https:// doi.org/10.1111/ciso.12464.

Elliott, Missy. *Supa Dupa Fly*. Produced by Timbaland. The Goldmind; East West; Elektra. Released 1997.

Ellison, Treva Carrie. "Black Femme Praxis and the Promise of Black Gender." *Black Scholar* 49, no. 1 (2019): 6–16. https://doi.org/10.1080/00064246 .2019.1548055.

England, Sarah. "Reading the Dougla Body: Mixed-Race, Post-Race, and Other Narratives of What It Means to Be Mixed in Trinidad." *Latin American and Caribbean Ethnic Studies* 3, no. 1 (2008): 1–31. https://doi.org/10.1080 /17442220701865820.

Environmental Protection Agency. "Flow Control and Municipal Solid Waste." Last updated March 29, 2016. https://archive.epa.gov/epawaste/nonhaz /municipal/web/html/flowctrl.html.

Environmental Protection Agency. *Flow Controls and Municipal Solid Waste: Executive Summary*. March 1995. https://archive.epa.gov/epawaste/nonhaz /municipal/web/pdf/execsum.pdf.

Environmental Protection Agency. *Fourth Five-Year Review Report for Suffolk City Landfill Superfund Site: Suffolk, Virginia*. August 2014. https://semspub.epa .gov/work/03/2194867.pdf.

Environmental Protection Agency. *Municipal Solid Waste Generation, Recycling, and Disposal in the United States: Facts and Figures for 2012*. Washington, DC: Environmental Protection Agency, 2014. https://www.epa.gov/sites /default/files/2015-09/documents/2012_msw_fs.pdf.

Environmental Protection Agency. "Superfund: National Priorities List (NPL)." Last updated October 9, 2024. https://www.epa.gov/superfund /superfund-national-priorities-list-npl.

Faber, Daniel R., and Eric J. Krieg. "Unequal Exposure to Ecological Hazards: Environmental Injustices in the Commonwealth of Massachusetts." *Environmental Health Perspectives* 110, suppl. 2 (2002): 277–88. https://doi .org/10.1289/ehp.02110s2277.

Fabian, Johannes. *Time and the Other: How Anthropology Makes Its Object.* New York: Columbia University Press, 2002.

Fanon, Frantz. *Black Skin, White Masks.* Translated by Richard Philcox. New York: Grove Press, 2008.

Fanon, Frantz. *The Wretched of the Earth.* Translated by Richard Philcox. New York: Grove Press, 2004.

Federici, Silvia. *Caliban and the Witch: Women, the Body and Primitive Accumulation.* London: Penguin Books, 2021.

Ferdinand, Malcom. *Decolonial Ecology: Thinking from the Caribbean World.* Translated by Anthony Paul Smith. Critical South. Cambridge: Polity Press, 2022.

Ferguson, Roderick A. *Aberrations in Black: Toward a Queer of Color Critique.* Critical American Studies Series. Minneapolis: University of Minnesota Press, 2004.

Field, Corinne T., and LaKisha Michelle Simmons, eds. *The Global History of Black Girlhood.* Urbana: University of Illinois Press, 2022.

Finney, Carolyn. *Black Faces, White Spaces: Reimagining the Relationship of African Americans to the Great Outdoors.* Chapel Hill: University of North Carolina Press, 2014.

Fortun, Kim. *Advocacy After Bhopal: Environmentalism, Disaster, New Global Orders.* Chicago: University of Chicago Press, 2001.

Frazier, Chelsea M. "Troubling Ecology: Wangechi Mutu, Octavia Butler, and Black Feminist Interventions in Environmentalism." *Critical Ethnic Studies* 2, no. 1 (2016): 40. https://doi.org/10.5749/jcritethnstud.2.1.0040.

Fredericks, Rosalind. *Garbage Citizenship: Vital Infrastructures of Labor in Dakar, Senegal.* Durham, NC: Duke University Press, 2018.

Fuentes, Lorena. "'The Garbage of Society': Disposable Women and the Socio-Spatial Scripts of Femicide in Guatemala." *Antipode* 52, no. 6 (2020): 1667–87. https://doi.org/10.1111/anti.12669.

Fullilove, Mindy Thompson. *Root Shock: How Tearing Up City Neighborhoods Hurts America, and What We Can Do About It.* 2nd ed. New York: New Village Press, 2016.

Fullwiley, Duana. *The Encultured Gene: Sickle Cell Health Politics and Biological Difference in West Africa.* Princeton, NJ: Princeton University Press, 2011.

Gaye, Marvin. *Let's Get It On.* Produced by Marvin Gaye. Tamla Records. Released 1973.

Gidwani, Vinay, and Anant Maringanti. "The Waste-Value Dialectic: Lumpen Urbanization in Contemporary India." *Comparative Studies of South Asia, Africa and the Middle East* 36, no. 1 (2016): 112–33. https://doi.org/10.1215 /1089201x-3482159.

Gidwani, Vinay, and Rajyashree N. Reddy. "The Afterlives of 'Waste': Notes from India for a Minor History of Capitalist Surplus." *Antipode* 43, no. 5 (2011): 1625–58. https://doi.org/10.1111/j.1467-8330.2011.00902.x.

Gilbert, Steven G., and Bernard Weiss. "A Rationale for Lowering the Blood Lead Action Level from 10 to 2µg/dL." *NeruoToxicology* 27, no. 5 (2006): 693–701. https://doi.org/10.1016/j.neuro.2006.06.008.

Gilmore, Ruth Wilson. *Abolition Geography: Essays Towards Liberation*. Edited by Brenna Bhandar and Alberto Toscano. London: Verso, 2022.

Gilmore, Ruth Wilson. *Golden Gulag: Prisons, Surplus, Crisis, and Opposition in Globalizing California*. American Crossroads 21. Berkeley: University of California Press, 2007.

Gilroy, Paul. *The Black Atlantic: Modernity and Double Consciousness*. Cambridge, MA: Harvard University Press, 2003.

Glissant, Édouard. *Poetics of Relation*. Translated by Betsy Wing. Ann Arbor: University of Michigan Press, 1997.

Goeman, Mishuana R. "Disrupting a Settler-Colonial Grammar of Place: The Visual Memoir of Hulleah Tsinhnahjinnie." In *Theorizing Native Studies*, edited by Audra Simpson and Andrea Smith. Durham, NC: Duke University Press, 2014. https://doi.org/10.1215/9780822376613-010.

Goffe, Rachel. "Reproducing the Plot: Making Life in the Shadow of Premature Death." *Antipode*, February 2022, 1024–46. https://doi.org/10.1111/anti.12812.

Goffe, Tao Leigh. "Stolen Life, Stolen Time." *South Atlantic Quarterly* 121, no. 1 (2022): 109–30. https://doi.org/10.1215/00382876-9561573.

Gordon, Avery. *Ghostly Matters: Haunting and the Sociological Imagination*. Minneapolis: University of Minnesota Press, 2008.

Grande, Sandy. *Red Pedagogy: Native American Social and Political Thought*. Lanham, MD: Rowman and Littlefield, 2004.

Grant, Richard. "Deep in the Swamps, Archaeologists Are Finding How Fugitive Slaves Kept Their Freedom." *Smithsonian*, September 2016.

Gregory, Steven. *Black Corona: Race and the Politics of Place in an Urban Community*. Princeton Studies in Culture/Power/History. Princeton, NJ: Princeton University Press, 1998.

Gregory, Steven. "Race, Rubbish, and Resistance: Empowering Difference in Community Politics." *Cultural Anthropology* 8, no. 1 (1993): 24–48.

Grove, Richard. *Green Imperialism: Colonial Expansion, Tropical Island Edens and the Origins of Environmentalism, 1600–1860*. Studies in Environment and History. Cambridge, MA: Cambridge University Press, 1996.

Gumbs, Alexis Pauline. "Keyword: Mothering." *Parapraxis* 1 (Winter 2022). https://www.parapraxismagazine.com/articles/mothering.

Haley, Sarah. *No Mercy Here: "Gender, Punishment, and the Making of Jim Crow Modernity."* Chapel Hill: University of North Carolina Press, 2016. https://doi.org/10.5149/northcarolina/9781469627595.001.0001.

Hall, Stuart. "Race, Articulation, and Societies Structured in Dominance." In *Selected Writings on Race and Difference*, edited by Paul Gilroy and Ruth Wilson Gilmore. Durham, NC: Duke University Press, 2021.

Hammonds, Evelynn. "Black (W)holes and the Geometry of Black Female Sexuality." *Differences: A Journal of Feminist Cultural Studies* 6, no. 2–3 (1994): 126–45. https://doi.org/10.1215/10407391-6-2-3-126.

Hammonds, Evelynn. "Toward a Genealogy of Black Female Sexuality: The Problematic of Silence." In *Feminist Genealogies, Colonial Legacies, Democratic Futures*, edited by M. Jacqui Alexander and Chandra Talpade Mohanty. New York: Routledge, 2013. https://doi.org/10.4324/9780203724200.

Harding, Sandra G. *Sciences from Below: Feminisms, Postcolonialities, and Modernities*. Durham, NC: Duke University Press, 2008.

Hare, Nathan. "Black Ecology." In "Black Cities: Colonies or City States?," *Black Scholar* 1, no. 6 (1970): 2–8.

Harney, Stefano, and Fred Moten. *The Undercommons: Fugitive Planning and Black Study*. Wivenhoe: Minor Compositions, 2013.

Harper, Scott. "SPSA Landfill Expansion in Jeopardy." *Virginian-Pilot*, March 3, 2008.

Harper, Scott. "SPSA Says Chesapeake Suit Is Causing Financial Woe." *Virginian-Pilot*, August 8, 2019.

Harris, Cheryl I. "Whiteness as Property." *Harvard Law Review* 106, no. 8 (1993): 1707–91. https://doi.org/10.2307/1341787.

Hartman, Saidiya. "The Belly of the World: A Note on Black Women's Labors." *Souls* 18, no. 1 (2016): 166–73. https://doi.org/10.1080/10999949.2016.1162596.

Hartman, Saidiya. *Lose Your Mother: A Journey Along the Atlantic Slave Route*. New York: Farrar, Straus and Giroux, 2008.

Hartman, Saidiya. *Scenes of Subjection: Terror, Slavery, and Self-Making in Nineteenth-Century America*. New York: Oxford University Press, 1997.

Hartman, Saidiya. "Venus in Two Acts." *Small Axe: A Caribbean Journal of Criticism* 12, no. 2 (2008): 1–14. https://doi.org/10.1215/-12-2-1.

Hartman, Saidiya. *Wayward Lives, Beautiful Experiments: Intimate Histories of Riotous Black Girls, Troublesome Women, and Queer Radicals*. New York: W. W. Norton, 2020.

Hawkins, Gay. "Down the Drain: Shit and the Politics of Disturbance." In *Culture and Waste: The Creation and Destruction of Value*, edited by Gay Hawkins and Stephen Muecke. Lanham, MD: Rowman and Littlefield, 2003.

Hawkins, Gay. *The Ethics of Waste: How We Relate to Rubbish*. Lanham, MD: Rowman and Littlefield, 2006.

Hawkins, Gay. "Plastic Bags: Living with Rubbish." *International Journal of Cultural Studies* 4, no. 1 (2001): 5–23. https://doi.org/10.1177/136787790100400101.

Hawthorne, Camilla. "Black Matters Are Spatial Matters: Black Geographies for the Twenty-First Century." *Geography Compass* 13, no. 11 (2019). https://doi.org/10.1111/gec3.12468.

Hernandez, Jillian. *Aesthetics of Excess: The Art and Politics of Black and Latina Embodiment*. Durham, NC: Duke University Press, 2020.

Hernberg, Sven. "Lead Poisoning in a Historical Perspective." *American Journal of Industrial Medicine* 38, no. 3 (2000): 244–54. https://doi.org/10.1002/1097-0274(200009)38:3<244::AID-AJIM3>3.0.CO;2-F.

Heynen, Nik. "'A Plantation Can Be a Commons': Re-Earthing Sapelo Island Through Abolition Ecology: The 2018 Neil Smith Lecture." *Antipode* 53, no. 1 (2021): 95–114. https://doi.org/10.1111/anti.12631.

Heynen, Nik, and Megan Ybarra. "On Abolition Ecologies and Making 'Freedom as a Place.'" *Antipode* 53, no. 1 (2021): 21–35. https://doi.org/10.1111/anti.12666.

Hicks, Jonathan P. "Mayor Announces New Assault on Graffiti, Citing Its Toll on City." *New York Times*, November 17, 1994. https://www.nytimes.com/1994/11/17/nyregion/mayor-announces-new-assault-on-graffiti-citing-its-toll-on-city.html.

Higginbotham, Evelyn Brooks. *Righteous Discontent: The Women's Movement in the Black Baptist Church, 1880–1920*. Cambridge, MA: Harvard University Press, 2003.

Hird, Myra J. "Knowing Waste: Towards an Inhuman Epistemology." *Social Epistemology* 26, no. 3–4 (2012): 453–69. https://doi.org/10.1080/02691728.2012.727195.

Hird, Myra J., and James Wilkes. "Colonial Ideologies of Waste: Implications for Land and Life." In "Confronting Waste," special issue, *Europe Now* 27 (May 2019). https://www.europenowjournal.org/2019/05/06/colonial-ideologies-of-waste-implications-for-land-and-life/.

Hird, Myra J., and Alexander Zahara. "Arctic Wastes." In *Anthropocene Feminism*, edited by Richard Grusin and the Center for 21st Century Studies. Center for 21st Century Studies. Minneapolis: University of Minnesota Press, 2017.

hooks, bell. *Black Looks: Race and Representation*. Boston: South End Press, 1992.

hooks, bell. *Feminism Is for Everybody: Passionate Politics*. 2nd ed. New York: Routledge, 2015.

hooks, bell. *Feminist Theory: From Margin to Center*. 3rd ed. New York: Routledge, 2015.

hooks, bell. "The Oppositional Gaze: Black Female Spectators." In *Feminist Film Theory: A Reader*, edited by Sue Thornham. Edinburgh: Edinburgh University Press, 1993.

hooks, bell. "Sexism and Misogyny: Who Takes the Rap?" *End Magazine*, February 1994. http://challengingmalesupremacy.org/wp-content/uploads/2015/04/Misogyny-gangsta-rap-and-The-Piano-bell-hooks.pdf.

hooks, bell. *Teaching to Transgress: Education as the Practice of Freedom*. New York: Routledge, 1994.

hooks, bell. *We Real Cool: Black Men and Masculinity*. New York: Routledge, 2004.

Horton-Stallings, LaMonda. *A Dirty South Manifesto: Sexual Resistance and Imagination in the New South.* Oakland: University of California Press, 2020.

Hosbey, Justin, and J. T. Roane. "A Totally Different Form of Living: On the Legacies of Displacement and Marronage as Black Ecologies." *Southern Cultures* 27, no. 1 (2021): 68–73. https://doi.org/10.1353/scu.2021.0009.

Hunter, Margaret, and Kathleen Soto. "Women of Color in Hip Hop: The Pornographic Gaze." *Race, Gender and Class* 16, no. ½ (2009): 170–91.

Hurston, Zora Neale. *Barracoon.* Delhi, Mumbai: Grapevine India, 2022.

Hurston, Zora Neale. *Mules and Men.* New York: Harper Perennial, 2008.

Hurston, Zora Neale, Edwidge Danticat, and Henry Louis Gates. *Their Eyes Were Watching God.* 75th anniversary ed. New York: Harper Perennial Modern Classics, 2013.

Hutchins, Sarah. "Officials Declare Dismal Swamp Fire Out After 111 Days." *Virginian-Pilot*, November 23, 2011.

Iheka, Cajetan. *African Ecomedia: Network Forms, Planetary Politics.* Durham, NC: Duke University Press, 2021. https://doi.org/10.1515/9781478022046.

Isenberg, Nancy G. *White Trash: The 400-Year Untold History of Class in America.* New York: Penguin Books, 2017.

Jackson, Zakiyyah Iman. *Becoming Human: Matter and Meaning in an Antiblack World.* Sexual Cultures. New York: New York University Press, 2020.

Jackson, Zakiyyah Iman. "'Theorizing in a Void': Sublimity, Matter, and Physics in Black Feminist Poetics." *South Atlantic Quarterly* 117, no. 3 (2018): 617–48. https://doi.org/10.1215/00382876-6942195.

Jacobs, Harriet Ann. *Incidents in the Life of a Slave Girl.* Boston, 1861.

Jacobs-Huey, Lanita. "The Natives Are Gazing and Talking Back: Reviewing the Problematics of Positionality, Voice, and Accountability Among 'Native' Anthropologists." *American Anthropologist* 104, no. 3 (2002): 791–804. https://doi.org/10.1525/aa.2002.104.3.791.

James, C. L. R. *The Black Jacobins: Toussaint l'Ouverture and the San Domingo Revolution.* 2nd ed. New York: Vintage Books, 1989.

James, Joy. "The Architects of Abolitionism: George Jackson, Angela Davis, and the Deradicalization of Prison Struggles." Lecture. Center for the Study of Slavery and Justice, Brown University, April 8, 2019.

James, Joy. "The Captive Maternal and Abolitionism." *TOPIA: Canadian Journal of Cultural Studies* 43 (September 2021): 9–23. https://doi.org/10.3138/topia-43-002.

Jemisin, N. K. *The City We Became.* New York: Orbit, 2021.

Jemisin, N. K. *The Fifth Season.* New York: Orbit, 2015.

Jemisin, N. K. *The Inheritance Trilogy (The Hundred Thousand Kingdoms, the Broken Kingdoms, the Kingdom of Gods, the Awakened Kingdom).* New York: Orbit, 2014.

Jemisin, N. K. *The Obelisk Gate.* New York: Orbit, 2016.

Jemisin, N. K. *The Stone Sky.* New York: Orbit, 2017.

Jemisin, N. K. *The World We Make.* New York: Orbit, 2023.

Jobson, Ryan Cecil. "The Case for Letting Anthropology Burn: Sociocultural Anthropology in 2019." *American Anthropologist* 122, no. 2 (2020): 259–71. https://doi.org/10.1111/aman.13398.

Jones, Claudia. *An End to the Neglect of The Problems of The Negro Woman!* New York: National Women's Commission, CPUSA, 1949.

Jordan, June. *Living Room: New Poems.* New York: Thunder's Mouth Press, 1985.

Jordan, June. *Soldier: A Poet's Childhood.* New York: Basic Books, 2001.

Juday, Luke, and Hamilton Lombard. *Portsmouth Demographic Study.* Weldon Cooper Center for Public Service, University of Virginia, June 2015. https://www.portsmouthva.gov/DocumentCenter/View/751/Portsmouth-Demographic-Study-PDF?bidId=.

Kaba, Mariame. *We Do This 'til We Free Us: Abolitionist Organizing and Transforming Justice.* Edited by Tamara K. Nopper. Chicago: Haymarket Books, 2021.

Kaplan, Sara Clarke. "Love and Violence/Maternity and Death: Black Feminism and the Politics of Reading (Un)Representability." *Black Women, Gender and Families* 1, no. 1 (2007): 94–124.

Karera, Axelle. "Blackness and the Pitfalls of Anthropocene Ethics." *Critical Philosophy of Race* 7, no. 1 (2019): 32–56. https://doi.org/10.5325/critphilrace.7.1.0032.

Katz, Michael B. *The Undeserving Poor: America's Enduring Confrontation with Poverty.* 2nd ed. Oxford: Oxford University Press, 2013.

Keeling, Kara. *Queer Times, Black Futures.* New York: New York University Press, 2019.

Kelley, Robin D. G. *Freedom Dreams: The Black Radical Imagination.* 20th anniversary ed. Boston: Beacon Press, 2022.

Kelley, Robin D. G. "'We Are Not What We Seem': Rethinking Black Working-Class Opposition in the Jim Crow South." *Journal of American History* 80, no. 1 (1993): 75–112. https://doi.org/10.2307/2079698.

Kennedy, Greg, ed. *An Ontology of Trash: The Disposable and Its Problematic Nature.* Albany: State University of New York Press, 2007.

King, Ryan Scott, Marc Mauer, and Tracy Huling. "An Analysis of the Economics of Prison Siting in Rural Communities." *Criminology and Public Policy* 3, no. 3 (2004): 453–80. https://doi.org/10.1111/j.1745-9133.2004.tb00054.x.

King, Tiffany Lethabo. *The Black Shoals: Offshore Formations of Black and Native Studies.* Durham, NC: Duke University Press, 2019.

King, Tiffany Lethabo. "The Labor of (Re)Reading Plantation Landscapes Fungible(ly)." *Antipode* 48, no. 4 (2016): 1022–39. https://doi.org/10.1111/anti.12227.

Krom, M. D. "An Evaluation of the Concept of Assimilative Capacity as Applied to Marine Waters." *Ambio* 15, no. 4 (1986): 208–14.

Laporte, Dominique. *History of Shit.* Cambridge, MA: MIT Press, 2002.

Larkin, Brian. "The Politics and Poetics of Infrastructure." *Annual Review of Anthropology* 42 (2013): 327–43.

Latour, Bruno. *We Have Never Been Modern.* Cambridge, MA: Harvard University Press, 1994.

Lea, Tess, and Paul Pholeros. "This Is Not a Pipe: The Treacheries of Indigenous Housing." *Public Culture* 22, no. 1 (2010): 187–209. https://doi.org/10.1215/08992363-2009-021.

Leong, Diana. "The Mattering of Black Lives: Octavia Butler's Hyperempathy and the Promise of the New Materialisms." *Catalyst: Feminism, Theory, Technoscience* 2, no. 2 (2016): 1–35. https://doi.org/10.28968/cftt.v2i2.28799.

Lepawsky, Josh. "The Changing Geography of Global Trade in Electronic Discards: Time to Rethink the E-Waste Problem." *Geographical Journal* 181, no. 2 (2015): 147–59. https://doi.org/10.1111/geoj.12077.

Lepawsky, Josh, and Chris McNabb. "Mapping International Flows of Electronic Waste." *Canadian Geographies / Géographies Canadiennes* 54, no. 2 (2010): 177–95. https://doi.org/10.1111/j.1541-0064.2009.00279.x.

Lewis, Gail. "Once More with My Sistren: Black Feminism and the Challenge of Object Use." *Feminist Review* 126, no. 1 (2020): 1–18. https://doi.org/10.1177/0141778920944372.

Lewis, Oscar. "The Culture of Poverty." In *On Understanding Poverty: Perspectives from the Social Sciences*, edited by Daniel Patrick Moynihan, 187–200. New York: Basic Books, 1969.

Liboiron, Max. *Pollution Is Colonialism.* Durham, NC: Duke University Press, 2021.

Liboiron, Max, and Josh Lepawsky. *Discard Studies: Wasting, Systems, and Power.* Cambridge, MA: MIT Press, 2022.

Lipton, Eric. "Five States Challenge New York's Trash Plan." *Washington Post*, February 7, 1999. https://www.washingtonpost.com/archive/local/1999/02/07/five-states-challenge-new-yorks-trash-plan/6abd4987-b4f6-4cdf-bd24-bf66e4765975/.

Lipton, Eric, and R. H. Melton. "Tons More of N.Y. Trash Headed for Va." *Washington Post*, January 12, 1999. https://www.washingtonpost.com/wp-srv/local/longterm/trash/trash0112.htm?noredirect=on.

Locke, John. *The Second Treatise of Government and a Letter Concerning Toleration.* London, 1824.

Lorde, Audre. "Poetry Is Not a Luxury." In *Sister Outsider: Essays and Speeches*, 36–39. Trumansburg, NY: Crossing Press, 1984.

Lorde, Audre. *Sister Outsider: Essays and Speeches.* Trumansburg, NY: Crossing Press, 1984.

Lukezic, Craig. "Soils and Settlement Location in 18th Century Colonial Tidewater Virginia." *Historical Archaeology* 24, no. 1 (1990): 1–17.

Lutz, Catherine, and Jane Collins. "The Photograph as an Intersection of Gazes: The Example of National Geographic." *Visual Anthropology Review* 7, no. 1 (1991): 134–49. https://doi.org/10.1525/var.1991.7.1.134.

Lynch, Patrick. "State Warns Waste Authority About Leaks at Suffolk Landfill." *Daily Press*, February 15, 2003. https://www.dailypress.com/news/dp-xpm-20030215-2003-02-15-0302150167-story.html.

Lyons, Kristina M. *Vital Decomposition: Soil Practitioners and Life Politics.* Durham, NC: Duke University Press, 2020. https://doi.org/10.1215/9781478009207.

MacBride, Samantha. *Recycling Reconsidered: The Present Failure and Future Promise of Environmental Action in the United States.* Cambridge, MA: MIT Press, 2012.

Makar, A. B., K. E. McMartin, M. Palese, and T. R. Tephly. "Formate Assay in Body Fluids: Application in Methanol Poisoning." *Biochemical Medicine* 13, no. 2 (1975): 117–26. https://doi.org/10.1016/0006-2944(75)90147-7.

Malkki, Liisa H. *Purity and Exile: Violence, Memory, and National Cosmology Among Hutu Refugees in Tanzania.* Chicago: University of Chicago Press, 1995.

Malm, Andreas. *Fossil Capital: The Rise of Steam-Power and the Roots of Global Warming.* London: Verso, 2016.

Marable, Manning. *How Capitalism Underdeveloped Black America: Problems in Race, Political Economy, and Society.* Chicago: Haymarket Books, 2018.

Marx, Karl. *Capital: An Abridged Edition.* Edited by David McLellan. Oxford: Oxford University Press, 2008. https://doi.org/10.1093/owc/9780199535705.001.0001.

Marx, Karl, and Friedrich Engels. *Manifesto of the Communist Party.* 22nd printing of 100th anniversary ed. New York: International, 1979.

Masco, Joseph. *The Nuclear Borderlands: The Manhattan Project in Post–Cold War New Mexico.* Princeton, NJ: Princeton University Press, 2006.

Mattelart, Armand. *The Invention of Communication.* Minneapolis: University of Minnesota Press, 1996.

Mattelart, Armand. *Networking the World, 1794–2000.* Minneapolis: University of Minnesota Press, 2000.

Mbembe, Achille. *On the Postcolony.* Studies on the History of Society and Culture 41. Berkeley: University of California Press, 2001.

McCabe, Robert. "Have a Sinking Feeling About Your Home? Now You Can Check It Out." *Virginian-Pilot,* August 12, 2007. https://www.pilotonline.com/news/article_9485c7f6-5fcf-518d-a98b-acd43b75e3a0.html.

M'charek, Amade. "Salty: Traces of Migration, Death, and the Art of Paying Attention." "Democracy in Question." Special issue, *IWMpost* 126 (2020).

McHugh, Corinn. "Landfill Added to EPA Superfund List." *Suffolk News-Herald,* February 18, 1990. https://virginiachronicle.com/?a=d&d=SNH19900218.1.1&e=-------en-20--1--txt-txIN--------.

McKittrick, Katherine. *Dear Science and Other Stories.* Errantries. Durham, NC: Duke University Press, 2021.

McKittrick, Katherine. *Demonic Grounds: Black Women and the Cartographies of Struggle.* Minneapolis: University of Minnesota Press, 2006.

McKittrick, Katherine. "Mathematics Black Life." *Black Scholar* 44, no. 2 (2014): 16–28. https://doi.org/10.1080/00064246.2014.11413684.

McKittrick, Katherine. "On Plantations, Prisons, and a Black Sense of Place." *Social and Cultural Geography* 12, no. 8 (2011): 947–63. https://doi.org/10.1080/14649365.2011.624280.

McKittrick, Katherine. "Plantation Futures." *Small Axe: A Caribbean Journal of Criticism* 17, no. 3 (2013): 1–15. https://doi.org/10.1215/07990537-2378892.

McKittrick, Katherine, ed. *Sylvia Wynter: On Being Human as Praxis.* Durham, NC: Duke University Press, 2015.

McMillan, Uri. *Embodied Avatars: Genealogies of Black Feminist Art and Performance.* New York: New York University Press, 2015.

Melosi, Martin V. *Fresh Kills: A History of Consuming and Discarding in New York City.* New York: Columbia University Press, 2020.

Melosi, Martin V. *Garbage in the Cities: Refuse, Reform, and the Environment.* Rev. ed. Pittsburgh, PA: University of Pittsburgh Press, 2005.

Melosi, Martin V. *The Sanitary City: Environmental Services in Urban America from Colonial Times to the Present.* Abridged ed. Pittsburgh, PA: University of Pittsburgh Press, 2008.

Melton, R. H. "N.Y. Mayor's Trash Talk Riles Va." *Washington Post,* January 15, 1999. https://www.washingtonpost.com/wp-srv/local/virginia15.htm.

Messerschmidt, J. W. "'We Must Protect Our Southern Women': On Whiteness, Masculinities, and Lynching." In *Race, Gender, and Punishment: From Colonialism to the War on Terror,* edited by Mary Bosworth and Jeanne Flavin. New Brunswick, NJ: Rutgers University Press, 2007.

Millar, Kathleen M. *Reclaiming the Discarded: Life and Labor on Rio's Garbage Dump.* Durham, NC: Duke University Press, 2018.

Mills, Charles. "Black Trash." In *Faces of Environmental Racism: Confronting Issues of Global Justice,* edited by Laura Westra and Bill E. Lawson, 2nd ed. Lanham, MD: Rowman and Littlefield, 2001.

Mitchell, Timothy. *Carbon Democracy: Political Power in the Age of Oil.* London: Verso, 2013.

Mitchell, Timothy. *Rule of Experts: Egypt, Techno-Politics, Modernity.* Berkeley: University of California Press, 2002.

Mitman, Gregg. *Breathing Space: How Allergies Shape Our Lives and Landscapes.* New Haven, CT: Yale University Press, 2007.

Mohai, Paul, and Bunyan Bryant. "Environmental Injustice: Weighing Race and Class as Fators in the Distribution of Environmental Hazards." *University of Colorado Law Review* 63 (1992): 921.

Mol, Annemarie. *The Body Multiple: Ontology in Medical Practice.* Durham, NC: Duke University Press, 2002. https://doi.org/10.1215/9780822384151.

Monk, Thelonious. *Big Band and Quartet in Concert.* Recorded December 1963, Lincoln Center, New York City. Produced by Teo Macero. Columbia CS 8964. Released 1964.

Monk, Thelonious. *It's Monk's Time.* Produced by Teo Macero. Columbia CS 8984. Released 1964.

Monk, Thelonious. *Solo Monk.* Produced by Teo Macero. Columbia CS 9149. Released 1965.

Monteiro, Fabrice. "*The Prophecy*." Institute Artist. 2015. https://instituteartist.com /The-Prophecy-Fabrice-Monteiro.

Moraga, Cherríe, and Gloria Anzaldúa. *This Bridge Called My Back: Writings by Radical Women of Color*. 2nd ed. New York: Kitchen Table, 1983.

Moreton-Robinson, Aileen. *The White Possessive: Property, Power, and Indigenous Sovereignty*. Minneapolis: University of Minnesota Press, 2015.

Morgan, Jennifer L. *Laboring Women: Reproduction and Gender in New World Slavery*. Early American Studies. Philadelphia: University of Pennsylvania Press, 2004.

Morgensen, Scott Lauria. "The Biopolitics of Settler Colonialism: Right Here, Right Now." *Settler Colonial Studies* 1, no. 1 (2011): 52–76. https://doi.org /10.1080/2201473X.2011.10648801.

Morgensen, Scott Lauria. *Spaces Between Us: Queer Settler Colonialism and Indigenous Decolonization*. Minneapolis: University of Minnesota Press, 2011.

Morris, J. Brent. *Dismal Freedom: A History of the Maroons of the Great Dismal Swamp*. Chapel Hill: University of North Carolina Press, 2022.

Morris, Rosalind C., ed. Introduction to *Can the Subaltern Speak? Reflections on the History of an Idea*, 1–18. New York: Columbia University Press, 2010.

Morrison, Toni. *Beloved*. London: Vintage, 2022.

Morrison, Toni. *Playing in the Dark: Whiteness and the Literary Imagination*. New York: Vintage Books, 1993.

Moulton, Cyrus. "Wheelabrator Millbury Is Cited as a Top Polluter, Example of 'Environmental Apartheid.'" *Telegram and Gazette* (Worcester), June 9, 2019.

Moynihan, Daniel Patrick. *The Negro Family; the Case for National Action*. Office of Policy Planning and Research, US Department of Labor. Washington, DC: US Printing Office, 1965. https://catalog.hathitrust.org/Record /002490183.

Mrázek, Rudolf. *Engineers of Happy Land: Technology and Nationalism in a Colony*. Princeton, NJ: Princeton University Press, 2002.

Muhammad, Khalil Gibran. *The Condemnation of Blackness: Race, Crime, and the Making of Modern Urban America*. Cambridge, MA: Harvard University Press, 2019.

Müller, Simone M. "Toxic Commons: Toxic Global Inequality in the Age of the Anthropocene." *Environmental History* 26, no. 3 (2021): 444–50. https://doi .org/10.1093/envhis/emab014.006.

Mullin and Lonergan Associates, Inc. *Analysis of Impediments to Fair Housing Choice: Hampton Roads Region, Virginia*. June 2011. https://cms3.revize.com /revize/portsmouth/Uploads/2022%20Agency%20Plan%20Tabs/TAB%20 3%20Fair%20Housing.pdf.

Murphy, M. "Alterlife and Decolonial Chemical Relations." *Cultural Anthropology* 32, no. 4 (2017): 494–503. https://doi.org/10.14506/ca32.4.02.

Murphy, M. "Chemical Regimes of Living." *Environmental History* 13, no. 4 (2008): 695–703.

Murphy, M. *Seizing the Means of Reproduction: Entanglements of Feminism, Health, and Technoscience.* Experimental Futures. Durham, NC: Duke University Press, 2012.

Mutu, Wangechi. *Wangechi Mutu: A Fantastic Journey.* Exhibition at the Brooklyn Museum, 2014.

Nagle, Robin. *Picking Up: On the Streets and Behind the Trucks with the Sanitation Workers of New York City.* New York: Farrar, Straus and Giroux, 2014.

"Nansemond Indian Nation." *Encyclopedia Virginia.* December 7, 2020. https://encyclopediavirginia.org/entries/nansemond-tribe.

Nash, Jennifer Christine. *The Black Body in Ecstasy: Reading Race, Reading Pornography.* Durham, NC: Duke University Press, 2014. https://doi.org/10.1215/9780822377030.

"The National Priority List." n.d. Environmental Protection Agency (EPA).

Nelson, Robert K., LaDale Winling, Richard Marciano, and Nathan Connolly. "Mapping Inequality: Redlining in New Deal America." American Panorama. Bed-Stuy, Brooklyn, 2016. https://dsl.richmond.edu/panorama/redlining/#loc=12/40.654/-74.116&city=brooklyn-ny&area=D8.

New York City Department of City Planning. *NYC 2000: Population Growth and Race/Hispanic Composition.* New York City: Department of City Planning, 2001. https://www.nyc.gov/assets/planning/download/pdf/data-maps/nyc-population/census2000/nyc20001.pdf.

New York City Department of Sanitation. *A Summary of Sanitation Rules and Regulations.* June 2015. https://www.nyc.gov/assets/dsny/docs/about_DSNY-rules-and-regulations_0815.pdf.

Newell, Stephanie. *Histories of Dirt: Media and Urban Life in Colonial and Postcolonial Lagos.* Durham, NC: Duke University Press, 2020.

Newman, Andy. "Giuliani's Trash-for-Culture Deal Doesn't Play in Virginia." *New York Times,* January 16, 1999. https://www.nytimes.com/1999/01/16/nyregion/giuliani-s-trash-for-culture-deal-doesn-t-play-in-virginia.html.

Nichols, Robert. *Theft Is Property! Dispossession and Critical Theory.* Radical Américas. Durham, NC: Duke University Press, 2020.

Nir, Sarah Maslin. "New York City Fights Scavengers over a Treasure: Trash." *New York Times,* March 21, 2016. https://www.nytimes.com/2016/03/21/nyregion/new-york-city-fights-scavengers-over-a-treasure-trash.html.

Nixon, Rob. *Slow Violence and the Environmentalism of the Poor.* Cambridge, MA: Harvard University Press, 2011.

Nkrumah, Kwame. *Neo-Colonialism: The Last Stage of Imperialism.* London: Panaf, 2004. Originally published in 1965.

The Notorious B.I.G. *Ready to Die.* Produced by Sean "Puffy" Combs, Easy Mo Bee, Chucky Thompson, DJ Premier, and Lord Finesse. Bad Boy (US); Arista (UK). Released 1994.

Nyong'o, Tavia. *Afro-Fabulations: The Queer Drama of Black Life.* New York: New York University Press, 2019.

O'Connor, Alice. *Poverty Knowledge: Social Science, Social Policy, and the Poor in Twentieth-Century US History.* Princeton, NJ: Princeton University Press, 2002.

Oddo, Vanessa M., Castiel Chen Zhuang, Sarah B. Andrea, et al. "Changes in Precarious Employment in the United States: A Longitudinal Analysis." *Scandinavian Journal of Work, Environment and Health* 47, no. 3 (2021): 171–80. https://doi.org/10.5271/sjweh.3939.

Opperman, Romy. "The Need for a Black Feminist Climate Justice." *CR: The New Centennial Review* 22, no. 1 (2022): 59–93. https://doi.org/10.14321/crnewcentrevi.22.1.0059.

Osborne, T. "Security and Vitality: Drains, Liberalism and Power in the Nineteenth Century." In *Foucault and Political Reason: Liberalism, Neo-Liberalism and Rationalities of Government*, edited by Andrew Barry, T. Osborne, and Nicholas Rose. London: UCL Press, 1996.

Osman, Suleiman. *The Invention of Brownstone Brooklyn: Gentrification and the Search for Authenticity in Postwar New York.* Oxford: Oxford University Press, 2011.

Outka, Paul. *Race and Nature from Transcendentalism to the Harlem Renaissance.* New York: Palgrave Macmillan, 2008.

Oyěwùmí, Oyèrónkẹ́. *The Invention of Women: Making an African Sense of Western Gender Discourses.* Minneapolis: University of Minnesota Press, 1997.

Parramore, Thomas C. *Norfolk: The First Four Centuries.* With Peter C. Stewart and Tommy Bogger. Charlottesville: University Press of Virginia, 2000.

Patterson, Orlando. *Slavery and Social Death: A Comparative Study; With a New Preface.* Cambridge, MA: Harvard University Press, 2018.

Peisner, Lynn. "Va. Trash Imports Rise." *Waste 360*, August 1, 2004. https://www.waste360.com/industry-insights/va-trash-imports-rise.

Pellow, David N. *Garbage Wars: The Struggle for Environmental Justice in Chicago.* Cambridge, MA: MIT Press, 2004.

Petry, Ann. *The Street.* Boston: Houghton Mifflin, 1998.

Petryna, Adriana. *Life Exposed: Biological Citizens After Chernobyl.* Princeton, NJ: Princeton University Press, 2013.

Philip, Marlene NourbeSe. *Zong!* Edited by Setaey Adamu Boateng. Middletown, CT: Wesleyan University Press, 2011.

Pilcher, James, Liz Dufour, Sarah Taddeo, and Matthew Prensky. "Dream Home Nightmares: Ryan Homes Buyers Face Delays, Hassles as Repairs Lag." *USA Today*, October 30, 2019. https://www.cincinnati.com/in-depth/news/2019/10/31/ryan-homes-construction-building-warranty-claims/3929496002/.

Pinto, Samantha. "Black Feminist Literacies: Ungendering, Flesh, and Post-Spillers Epistemologies of Embodied and Emotional Justice." *Journal of Black Sexuality and Relationships* 4, no. 1 (2017): 25–45. https://doi.org/10.1353/bsr.2017.0019.

Pough, Gwendolyn D. "What It Do, Shorty? Women, Hip-Hop, and a Feminist Agenda." *Black Women, Gender and Families* 1, no. 2 (2007): 78–99.

Prasad, Indulata. "Towards Dalit Ecologies." *Environment and Society* 13, no. 1 (2022): 98–120. https://doi.org/10.3167/ares.2022.130107.

Preston, Alan. "Assimilative Capacity of U.S. Coastal Waters for Pollutants (Proceedings of a Workshop at Crystal Mountain, Washington, July 29–August 4, 1979) [. . .]." *Environmental Conservation* 7, no. 4 (1980): 338–39. https://doi.org/10.1017/S0376892900008316.

Pristin, Terry. "In Bedford-Stuyvesant, the Boom Remains a Bust; Neighborhood Development Unit Under Fire." *New York Times*, May 29, 2000. https://www.nytimes.com/2000/05/29/nyregion/bedford-stuyvesant-boom-remains-bust-neighborhood-development-unit-under-fire.html.

Puar, Jasbir K. *The Right to Maim: Debility, Capacity, Disability.* Durham, NC: Duke University Press, 2017.

Pulido, Laura. "Flint, Environmental Racism and Racial Capitalism." *Capitalism Nature Socialism* 27, no. 3 (2016): 1–16. https://doi.org/10.1080/10455752.2016.1213013.

Pulido, Laura. "Geographies of Race and Ethnicity II: Environmental Racism, Racial Capitalism and State-Sanctioned Violence." *Progress in Human Geography* 41, no. 4 (2017): 524–33. https://doi.org/10.1177/0309132516646495.

Pulido, Laura. "Rethinking Environmental Racism: White Privilege and Urban Development in Southern California." In *Critical Geographies: A Collection of Readings*, edited by Harald Bauder and Salvatore Engel-Di Mauro, 532–577. Praxis (e)Press, 2008.

Purnell, Brian. *Fighting Jim Crow in the County of Kings: The Congress of Racial Equality in Brooklyn.* Lexington: University Press of Kentucky, 2015.

Purnell, Brian. "Taxation Without Sanitation Is Tyranny." In *Civil Rights in New York City: From World War II to the Giuliani Era*, edited by Clarence Taylor. New York: Fordham University Press, 2011. https://doi.org/10.1515/9780823237463.

Quashie, Kevin Everod. *The Sovereignty of Quiet: Beyond Resistance in Black Culture.* New Brunswick, NJ: Rutgers University Press, 2012.

Rainey, Ma. "See, See Rider Blues." Recorded October 16, 1924. Paramount 12252-B. Released 1925.

Rathje, William L., and Cullen Murphy. *Rubbish! The Archaeology of Garbage.* Tucson: University of Arizona Press, 2001.

Reddy, Rajyashree N. "Reimagining E-Waste Circuits: Calculation, Mobile Policies, and the Move to Urban Mining in Global South Cities." *Urban Geography* 37, no. 1 (2016): 57–76. https://doi.org/10.1080/02723638.2015.1046710.

Redfield, Peter. *Space in the Tropics: From Convicts to Rockets in French Guiana.* Berkeley: University of California Press, 2000.

Reese, Ashanté M. *Black Food Geographies: Race, Self-Reliance, and Food Access in Washington, D.C.* Chapel Hill: University of North Carolina Press, 2019.

Regional Solid Waste Disposal System: Annual Survey Report, Southeastern Public Service Authority of Virginia. Suffolk, VA: Louis Berger, 2015. https://www.spsa.com/application/files/7115/7101/3203/2015-Annual-Survey-and-Report-040115.pdf.

Regis, Ferne Louanne. "The Dougla in Trinidad's Consciousness." *History in Action* 2, no. 1 (2011).

Reno, Joshua. *Waste Away: Working and Living with a North American Landfill.* Oakland: University of California Press, 2016.

Resnick, Elana. "The Limits of Resilience: Managing Waste in the Racialized Anthropocene." *American Anthropologist* 123, no. 2 (2021): 222–36. https://doi.org/10.1111/aman.13542.

"Resolutions of the Dismal Swamp Company, 1 May 1785." George Washington Papers, Series 9, Addenda, 1763–1797. Library of Congress, Washington, DC.

Roane, J. T. "Black Ecologies, Subaquatic Life, and the Jim Crow Enclosure of the Tidewater." *Journal of Rural Studies* 94 (2022): 227–38. https://doi.org/10.1016/j.jrurstud.2022.06.006.

Roane, J. T. *Dark Agoras: Insurgent Black Social Life and the Politics of Place.* New York: New York University Press, 2023. https://doi.org/10.18574/nyu/9781479845385.001.0001.

Roane, J. T. "Plotting the Black Commons." *Souls* 20, no. 3 (2018): 239–66. https://doi.org/10.1080/10999949.2018.1532757.

Roberts, Dorothy E. *Killing the Black Body: Race, Reproduction, and the Meaning of Liberty.* 2nd ed. New York: Vintage Books, 2017.

Roberts, Dorothy E. "*Loving v. Virginia* as a Civil Rights Decision." *New York Law School Law Review* 59, no. 1 (2014): 175–209.

Robinson, Cedric J. *Black Marxism: The Making of the Black Radical Tradition.* 3rd ed. Chapel Hill: University of North Carolina Press, 2020.

Rosengren, Cole. "Update: Lechate Drainage Efforts More Efficient Than Expected at Suffolk, VA Landfill." *Waste Dive*, May 25, 2017. https://www.wastedive.com/news/update-leachate-drainage-efforts-more-efficient-than-expected-at-suffolk/439802/.

Rothstein, Richard. *The Color of Law: A Forgotten History of How Our Government Segregated America.* New York: Liveright, 2017.

Rowe, Aimee Carrillo, and Eve Tuck. "Settler Colonialism and Cultural Studies: Ongoing Settlement, Cultural Production, and Resistance." *Cultural Studies ↔ Critical Methodologies* 17, no. 1 (2017): 3–13. https://doi.org/10.1177/1532708616653693.

Royster, Deirdre Alexia. *Race and the Invisible Hand: How White Networks Exclude Black Men from Blue-Collar Jobs.* Berkeley: University of California Press, 2003.

Rushin, Donna Kate. "The Bridge Poem." In *This Bridge Called My Back: Writings by Radical Women of Color*, 2nd ed., edited by Cherríe Moraga and Gloria Anzaldúa. New York: Kitchen Table, 1983.

Ryan, William. *Blaming the Victim*. New York: Random House, 1988.

Sandoval, Chela, and Angela Y. Davis. *Methodology of the Oppressed*. Minneapolis: University of Minnesota Press, 2008.

Sassen, Saskia. *Expulsions: Brutality and Complexity in the Global Economy*. Cambridge, MA: Belknap Press of Harvard University Press, 2014.

Sassen, Saskia. *The Global City: New York, London, Tokyo*. 2nd ed. Princeton, NJ: Princeton University Press, 2001.

Sayers, Daniel O. *A Desolate Place for a Defiant People: The Archaeology of Maroons, Indigenous Americans, and Enslaved Laborers in the Great Dismal Swamp*. Gainesville: University Press of Florida, 2015.

Schalk, Sami. *Black Disability Politics*. Durham, NC: Duke University Press, 2022.

Schept, Judah. "(Un)seeing Like a Prison: Counter-Visual Ethnography of the Carceral State." *Theoretical Criminology* 18, no. 2 (2014): 198–223. https://doi.org/10.1177/1362480613517256.

Schmaltz, Jeff. "Fire in Great Dismal Swamp, Virginia." NASA, 2011.

Schuller, Kyla. *The Biopolitics of Feeling: Race, Sex, and Science in the Nineteenth Century*. Durham, NC: Duke University Press, 2018.

Scott, James C. *Seeing Like a State: How Certain Schemes to Improve the Human Condition Have Failed*. New Haven, CT: Yale University Press, 2020.

SCS Engineers. *Remedial Investigation Report for the Hosier Road Landfill, Suffolk, Virginia*. August 1992. https://semspub.epa.gov/work/03/138003.pdf.

Sedgwick, Eve Kosofsky. "Paranoid Reading and Reparative Reading, or, You're so Paranoid, You Probably Think This Essay Is About You." In *Touching Feeling: Affect, Pedagogy, Performativity*. Durham, NC: Duke University Press, 2003.

Sexton, Jared. "The Social Life of Social Death." *InTensions* 5, no. 1 (2011): 1–47.

Shabazz, Rashad. *Spatializing Blackness: Architectures of Confinement and Black Masculinity in Chicago*. Urbana: University of Illinois Press, 2015.

Shabazz, Rashad. "'We Gon Be Alright': Containment, Creativity, and the Birth of Hip-Hop." *Cultural Geographies* 28, no. 3 (2021): 447–53. https://doi.org/10.1177/14744740211003653.

Shalhoub-Kevorkian. "Infiltrated Intimacies: The Case of Palestinian Returnees." *Feminist Studies* 42, no. 1 (2016): 166. https://doi.org/10.15767/feministstudies.42.1.166.

Shange, Savannah. *Progressive Dystopia: Abolition, Antiblackness, and Schooling in San Francisco*. Durham, NC: Duke University Press, 2019.

Sharp, Lesley Alexandra. *Strange Harvest: Organ Transplants, Denatured Bodies, and the Transformed Self*. Berkeley: University of California Press, 2006.

Sharpe, Christina Elizabeth. *In the Wake: On Blackness and Being*. Durham, NC: Duke University Press, 2016.

Sharpe, Christina Elizabeth. *Monstrous Intimacies: Making Post-Slavery Subjects.* Durham, NC: Duke University Press, 2010.

Shaw, Ian G. R., and Marv Waterstone. "A Planet of Surplus Life: Building Worlds Beyond Capitalism." *Antipode* 53, no. 6 (2021): 1787–1806.

Shaw, Ian G. R., and Marv Waterstone. *Wageless Life: A Manifesto for a Future Beyond Capitalism.* Minneapolis: University of Minnesota Press, 2019.

Shotwell, Alexis. *Against Purity: Living Ethically in Compromised Times.* Minneapolis: University of Minnesota Press, 2016.

Silkenat, David. *Scars on the Land: An Environmental History of Slavery in the American South.* New York: Oxford University Press, 2022.

Simmons, Kristen. "Settler Atmospherics." *Society for Cultural Anthropology: Fieldsights* (blog). November 20, 2017.

Simmons, LaKisha Michelle. "Black Girlhood in the Nineteenth Century by Nazera Sadiq Wright." *Journal of the History of Childhood and Youth* 11, no. 1 (2018): 124–26. https://doi.org/10.1353/hcy.2018.0017.

Simone, AbdouMaliq. *Improvised Lives.* Cambridge: Polity Press, 2019.

Simone, AbdouMaliq. "People as Infrastructure: Intersecting Fragments in Johannesburg." *Public Culture* 16, no. 3 (2004): 407–29. https://doi.org/10.1215/08992363-16-3-407.

Simone, AbdouMaliq. "To Extend: Temporariness in a World of Itineraries." *Urban Studies* 57, no. 6 (2020): 1127–42. https://doi.org/10.1177/0042098020905442.

Simpson, Audra. *Mohawk Interruptus: Political Life Across the Borders of Settler States.* Durham, NC: Duke University Press, 2014. https://doi.org/10.1215/9780822376781.

Simpson, Audra. "The State Is a Man: Theresa Spence, Loretta Saunders and the Gender of Settler Sovereignty." *Theory and Event* 19, no. 4 (2016).

Simpson, Elizabeth. "Lead Poisoning: It's a Problem That Hits Close to Home." *Virginian-Pilot*, December 17, 2007. https://www.pilotonline.com/news/article_6f9405c1-2291-57a9-928b-95cf4b20c550.html.

Smith, Neil. "Giuliani Time: The Revanchist 1990s." *Social Text* 57, no. (1998): 1–20. https://doi.org/10.2307/466878.

Smith, Neil. *Uneven Development: Nature, Capital, and the Production of Space.* 3rd ed. Athens: University of Georgia Press, 2008.

Snelgrove, Corey, Rita Kaur Dhamoon, and Jeff Corntassel. "Unsettling Settler Colonialism: The Discourse and Politics of Settlers, and Solidarity with Indigenous Nations." *Decolonization: Indigeneity, Education and Society* 3, no. 2 (2014): 1–32.

Snorton, C. Riley. *Black on Both Sides: A Racial History of Trans Identity.* Minneapolis: University of Minnesota Press, 2017.

Snorton, C. Riley. "Referential Sights and Slights." *Palimpsest: A Journal on Women, Gender, and the Black International* 2, no. 2 (2013): 175–86. https://doi.org/10.1353/pal.2013.0021.

Snorton, C. Riley, and Jin Haritaworn. "Trans Necropolitics: A Transnational
    Reflection on Violence, Death, and the Trans of Color Afterlife." In *The
    Transgender Studies Reader 2*, edited by Susan Stryker and Aren Z. Aizura.
    New York: Routledge, 2013.

Solomon, Marisa. "'The Ghetto Is a Gold Mine': The Racialized Temporality of Bet-
    terment." *International Labor and Working-Class History* 95 (2019): 76–94.
    https://doi.org/10.1017/S0147547919000024.

Solomon, Marisa, and Zoë Wool. "Waste Is Not a Metaphor for Racist Disposses-
    sion: The Black Feminist Marxism of Marisa Solomon." *Catalyst: Femi-
    nism, Theory, Technoscience* 7, no. 2 (2021): 1–5. https://doi.org/10.28968
    /cftt.v7i2.37655.

Spade, Dean. *Normal Life: Administrative Violence, Critical Trans Politics, and the Limits
    of Law.* Rev. ed. Durham, NC: Duke University Press, 2015.

Speiran, Gary K., and Frederic C. Wurster. *Hydrology and Water Quality of the Great
    Dismal Swamp, Virginia and North Carolina, and Implications for Hydrologic-
    Management Goals and Strategies.* US Geological Survey Scientific Inves-
    tigations Report 2020–5100, 2021. https://pubs.usgs.gov/publication
    /sir20205100#:~:text=Timber%20harvesting%20and%20the%20
    construction,increased%20the%20risk%20of%20fire.

Spillers, Hortense J. "Mama's Baby, Papa's Maybe: An American Grammar Book."
    Diacritics 17, no. 2 (1987): 64–81. https://doi.org/10.2307/464747.

Spivak, Gayatri Chakravorty. "In Response: Looking Back, Looking Forward."
    In *Can the Subaltern Speak? Reflections on the History of an Idea*, ed-
    ited by Rosalind C. Morris. New York: Columbia University Press,
    2010.

Spivak, Gayatri Chakravorty. "Can the Subaltern Speak?" In *Can the Subaltern
    Speak? Reflections on the History of an Idea*, edited by Rosalind C. Morris.
    New York: Columbia University Press, 2010.

Stallings, L. H. "Hip Hop and the Black Ratchet Imagination." *Palimpsest: A Journal
    on Women, Gender, and the Black International* 2, no. 2 (2013): 135–39.
    https://doi.org/10.1353/pal.2013.0026.

Stamatopoulou-Robbins, Sophia. *Waste Siege: The Life of Infrastructure in Palestine.*
    Stanford, CA: Stanford University Press, 2020.

Staples, Brent. "What If the Court in the Loving Case Had Declared Race a False
    Idea?" *New York Times*, March 6, 2017.

Star, Susan Leigh. "The Ethnography of Infrastructure." *American Behav-
    ioral Scientist* 43, no. 3 (1999): 377–91. https://doi.org/10.1177/00027
    649921955326.

Steinberg, Theodore. *Gotham Unbound: The Ecological History of Greater New York.*
    New York: Simon and Schuster, 2015.

Stewart, Kathleen. *Ordinary Affects.* Durham, NC: Duke University Press, 2007.

Stoler, Ann Laura, ed. *Imperial Debris: On Ruins and Ruination.* Durham, NC: Duke
    University Press, 2013.

Strach, Patricia, and Kathleen S. Sullivan. *The Politics of Trash: How Governments Used Corruption to Clean Cities, 1890–1929.* Ithaca, NY: Cornell University Press, 2022.

Strasser, Susan. *Waste and Want: A Social History of Trash.* New York: Henry Holt, 2000.

Strings, Sabrina. *Fearing the Black Body: The Racial Origins of Fat Phobia.* New York: New York University Press, 2019.

Strings, Sabrina. "Obese Black Women as 'Social Dead Weight': Reinventing the "Diseased Black Woman." *Signs: Journal of Women in Culture and Society* 41, no. 1 (2015): 107–30. https://doi.org/10.1086/681773.

Summers, Brandi Thompson. *Black in Place: The Spatial Aesthetics of Race in a Post-Chocolate City.* Chapel Hill: University of North Carolina Press, 2019.

Swiffen, Amy, and Shoshana Paget. 2022. "The Biopolitics of Settler Colonialism and the Limits of Foucault's Historical Method." In *Studies in Law, Politics, and Society*, edited by Austin Sarat, George Pavlich, and Richard Mailey. Emerald Publishing Limited. https://doi.org/10.1108/S1059 -43372022000087A006.

Sze, Julie. *Environmental Justice in a Moment of Danger.* Oakland: University of California Press, 2020.

Sze, Julie. *Noxious New York: The Racial Politics of Urban Health and Environmental Justice.* Cambridge, MA: MIT Press, 2007.

Tadiar, Neferti X. M. *Remaindered Life.* Durham, NC: Duke University Press, 2022.

TallBear, Kim. *Native American DNA: Tribal Belonging and the False Promise of Genetic Science.* Minneapolis: University of Minnesota Press, 2013.

Taylor, Clarence. *Fight the Power: African Americans and the Long History of Police Brutality in New York City.* New York: New York University Press, 2019.

Taylor, Dorceta E. *Toxic Communities: Environmental Racism, Industrial Pollution, and Residential Mobility.* New York: New York University Press, 2014.

Taylor, Keeanga-Yamahtta, ed. "Combahee River Collective Statement." In *How We Get Free: Black Feminism and the Combahee River Collective.* Chicago: Haymarket Books, 2017.

Taylor, Keeanga-Yamahtta, ed. *Race for Profit: How Banks and the Real Estate Industry Undermined Black Homeownership.* Chapel Hill: University of North Carolina Press, 2019.

Thomson, Vivian E. *Garbage In, Garbage Out: Solving the Problems with Long-Distance Trash Transport.* Charlottesville: University of Virginia Press, 2009.

Thornham, Sue, ed. *Feminist Film Theory: A Reader.* Repr. Edinburgh: Edinburgh University Press, 2005.

Ticktin, Miriam. "Migrant Occupations." Forthcoming.

Toxic Wastes and Race in the United States: A National Report on the Racial and Socio- Economic Characteristics of Communities with Hazardous Waste Sites. New

York: United Church of Christ, Commission for Racial Justice, 1987. https://d3n8a8pro7vhmx.cloudfront.net/unitedchurchofchrist/legacy _url/13567/toxwrace87.pdf.

Trask, Haunani-Kay. "The Color of Violence." In *Color of Violence: The INCITE! Anthology*, edited by INCITE! Women of Color Against Violence. Durham, NC: Duke University Press, 2016.

Trouillot, Michel-Rolph. *Silencing the Past: Power and the Production of History.* Boston: Beacon Press, 1995.

Tso, T. C. *Physiology and Biochemistry of Tobacco Plants.* Stroudsburg, PA: Dowden Hutchenson and Ross, 1974.

Tuck, Eve, Mistinguette Smith, Allison M. Guess, Tavia Benjamin, and Brian K. Jones. "Geotheorizing Black/Land." *Departures in Critical Qualitative Research* 3, no. 1 (2014): 52–74. https://doi.org/10.1525/dcqr.2014.3.1.52.

Tuck, Eve, and K. Wayne Yang. "Decolonization Is Not a Metaphor." *Decolonization: Indigeneity, Education and Society* 1, no. 1 (2012): 1–40.

Tuck, Eve, and K. Wayne Yang. "R-Words: Refusing Research." In *Humanizing Research: Decolonizing Qualitative Inquiry with Youth and Communities*, edited by Django Paris and Maisha T. Winn. Los Angeles: Sage, 2014. https://doi.org/10.4135/9781544329611.

Tumarkin, Laurel, and Jonathan Bowles. "Staten Island: Then and Now." New York City: Staten Island Economic Development Corporation, 2011. https://nycfuture.org/pdf/Staten_Island_-_Then_and_Now.pdf.

Turken, Sam. "'It Was Horrific': Norfolk's Predominantly Black East Ghent Wiped Off the Map in 1970s." *WHRO Public Media*, July 21, 2020. https://www.whro.org/2020-07-21/it-was-horrific-norfolk-s-predominantly-black -east-ghent-wiped-off-the-map-in-1970s.

UN Environment Programme. "How Tweaking Your Diet Can Help Save the Planet." *UNEP News and Stories* (blog). August 23, 2021. https://www.unep.org /news-and-stories/story/how-tweaking-your-diet-can-help-save-planet.

Urban Renewal Center. "Changing Tides: Gentrification in Norfolk." Online presentation, Norfolk, VA, August 22, 2019. https://theurcnorfolk.com /research-gentrification.

US Department of Agriculture. *Soil Survey of City of Suffolk, Virginia.* US Department of Agriculture Soil Conservation Service, June 1981. https://www.nrcs.usda.gov/Internet/FSE_MANUSCRIPTS/virginia /cityofsuffolkVA1981/suffolkVA1981.pdf.

US Department of Energy, Office of Scientific and Technical Information. "Health Assessment for Suffolk City Landfill, Suffolk, Nansemond County, Virginia, Region 3. CERCLIS No. VAD980917983. Preliminary Report." Agency for Toxic Substances and Disease Registry, 1991. https://www.osti.gov/biblio/5606811.

US Fish and Wildlife Service. "Resistance and Refuge." Interpretive panels, the Great Dismal Swamp, Suffolk, Virginia, 2012.

Uteuova, Aliya. "After Slavery, Oystering Offered a Lifeline. Now Sewage Spills Threaten to End It All." *Guardian*, September 1, 2021. https://www .theguardian.com/us-news/2021/sep/01/sewage-spills-crisis-virginia -oysters-black-communities.

Vasudevan, Pavithra. "An Intimate Inventory of Race and Waste." *Antipode* 53, no. 3 (2021): 770–90. https://doi.org/10.1111/anti.12501.

Vergès, Françoise. "Capitalocene, Waste, Race, and Gender." *E-flux* 100 (May 2019). https://www.e-flux.com/journal/100/269165/capitalocene-waste -race-and-gender/.

Vergès, Françoise. "Racial Capitalocene." In *Future of Black Radicalism*, edited by Gaye Theresa Johnson and Alex Lubin. London: Verso, 2017.

Virginia Asthma Coalition. *Virginia Asthma Plan, 2011–2016*. Virginia Asthma Co-alition, August 2010. https://archive.epa.gov/reg3artd/archive/web/pdf /asthmaplan.8.30.10.pdf.

Virginia Department of Environmental Quality. *Annual Solid Waste Report for CY2022*. Commonwealth of Virginia, October 2023.

Virginia Department of Health. *Virginia Asthma Burden Report*, 2018. https://www .vdh.virginia.gov/content/uploads/sites/94/2018/11/Asthma-Burden -Report_Final_10232018-1.pdf.

Virginia Department of Health and Portsmouth Health District. *The 2017 Portsmouth Community Health Survey (CHS) Using CASPER Methodology*, October 2018. https://www.ghrconnects.org/content/sites/uwshr /Community_Initiative_Pages/Healthy_Portsmouth/Resources/CHS -2017-Report-FINAL.pdf.

Virginia Department of Transportation. *A History of Roads in Virginia: "The Most Convenient Wayes."* Office of Public Affairs, Virginia Department of Transportation, 2006. https://www.virginiadot.org/about/resources /historyofrds.pdf.

Virginian-Pilot. "Landfill Solution Is 500 Feet Short." March 9, 2008. https://www .pilotonline.com/2008/03/09/landfill-solution-is-500-feet-short/.

Visperas, Cristina Mejia. *Skin Theory: Visual Culture and the Postwar Prison Laboratory*. New York: New York University Press, 2022. https://doi.org/10.18574 /nyu/9781479810772.001.0001.

Visweswaran, Kamala. *Fictions of Feminist Ethnography*. Minneapolis: University of Minnesota Press, 2003.

Von Schnitzler, Antina. *Democracy's Infrastructure: Techno-Politics and Protest After Apartheid*. Princeton, NJ: Princeton University Press, 2016.

Wacquant, Loïc J. D. *Punishing the Poor: The Neoliberal Government of Social Insecurity*. Durham, NC: Duke University Press, 2009.

Walcott, Rinaldo. *On Property*. Windsor, Ontario: Biblioasis, 2021.

Walker, Ameena. "The Notorious B.I.G. Mural in Bed-Stuy Will Come Down." *New York Curbed*, May 19, 2017. https://ny.curbed.com/2017/5/19/15662140 /brooklyn-biggie-smalls-mural-destroyed.

Warren, Calvin L. *Ontological Terror: Blackness, Nihilism, and Emancipation,* Durham, NC: Duke University Press, 2018.

Weheliye, Alexander G. *Habeas Viscus: Racializing Assemblages, Biopolitics, and Black Feminist Theories of the Human.* Durham, NC: Duke University Press, 2014.

Weinbaum, Alys Eve. "Gendering the General Strike: W. E. B. Du Bois's *Black Reconstruction* and Black Feminism's 'Propaganda of History.'" *South Atlantic Quarterly* , no. 3 (2013): 437–63. https://doi.org/10.1215/00382876-2146395.

Westra, Laura, and Bill E. Lawson, eds. "Decision Making." In *Faces of Environmental Racism: Confronting Issues of Global Justice,* 2nd ed. Lanham, MD: Rowman and Littlefield, 2001.

Whalley, Elizabeth, and Colleen Hackett. "Carceral Feminisms: The Abolitionist Project and Undoing Dominant Feminisms." *Contemporary Justice Review* 20, no. 4 (2017): 456–73. https://doi.org/10.1080/10282580.2017.1383762.

White, Forrest R. *Black, White and Brown: The Battle for Progress in 1950s Norfolk.* Norfolk, VA: Parke Press, 2018.

White, Theresa Renee. "Missy 'Misdemeanor' Elliott and Nicki Minaj: Fashionistin' Black Female Sexuality in Hip-Hop Culture—Girl Power or Overpowered?" *Journal of Black Studies* 44, no. 6 (2013): 607–26.

Whyte, Kyle. "Settler Colonialism, Ecology, and Environmental Injustice." *Environment and Society* 9, no. 1 (2018): 125–44. https://doi.org/10.3167/ares.2018.090109.

Wideman, Trevor J. "Property, Waste, and the 'Unnecessary Hardship' of Land Use Planning in Winnipeg, Canada." *Urban Geography* 41, no. 6 (2020): 865–92. https://doi.org/10.1080/02723638.2019.1698866.

Wilder, Craig Steven. *A Covenant with Color: Race and Social Power in Brooklyn.* New York: Columbia University Press, 2000.

Wilderson, Frank B. *Red, White and Black: Cinema and the Structure of U.S. Antagonisms.* Durham, NC: Duke University Press, 2010.

Williams, Brian, and Jayson Maurice Porter. "Cotton, Whiteness, and Other Poisons." *Environmental Humanities* 14, no. 3 (2022): 499–521. https://doi.org/10.1215/22011919-9962827.

Williams, James Gordon. *Crossing Bar Lines: The Politics and Practices of Black Musical Space.* Jackson: University Press of Mississippi, 2021.

Williams, Teona. "For 'Peace, Quiet, and Respect': Race, Policing, and Land Grabbing on Chicago's South Side: The 2018 Clyde Woods Black Geographies Specialty Group Graduate Student Paper Award." *Antipode* 53, no. 2 (2021): 497–523. https://doi.org/10.1111/anti.12692.

Williamson, Terrion L. *Scandalize My Name: Black Feminist Practice and the Making of Black Social Life.* New York: Fordham University Press, 2017.

Winner, Langdon. "Do Artifacts Have Politics?" *Daedalus* 109, no. 1 (1980): 121–36.

Wolfe, Patrick. "Settler Colonialism and the Elimination of the Native." *Journal of Genocide Research* 8, no. 4 (2006): 387–409. https://doi.org/10.1080/14623520601056240.

Woods, Clyde Adrian. *Development Arrested: The Blues and Plantation Power in the Mississippi Delta*. London: Verso, 2017.

Woodsworth, Michael. *Battle for Bed-Stuy: The Long War on Poverty in New York City*. Cambridge, MA: Harvard University Press, 2016.

Wright, Richard. *Native Son*. London: Jonathan Cape, 1970.

Wright, Richard. *Native Son*. Princeton, NJ: Recording for the Blind and Dyslexic, 2008.

Wright, Willie Jamaal. "As Above, So Below: Anti-Black Violence as Environmental Racism." *Antipode* 53, no. 3 (2021): 791–809. https://doi.org/10.1111/anti .12425.

Wright, Willie Jamaal. "The Morphology of Marronage." *Annals of the American Association of Geographers* 110, no. 4 (2020): 1134–49. https://doi.org/10.1080 /24694452.2019.1664890.

Wynter, Sylvia. "From 'Beyond Miranda's Meanings: Un/Silencing the "Demonic Ground" of Caliban's "Woman." '" In *Post-Colonial Theory and English Literature: A Reader*, edited by Peter Child. Edinburgh: Edinburgh University Press, 1999.

Wynter, Sylvia. "Unsettling the Coloniality of Being/Power/Truth/Freedom: Towards the Human, After Man, Its Overrepresentation—an Argument." *CR: The New Centennial Review* 3, no. 3 (2003): 257–337. https://doi .org/10.1353/ncr.2004.0015.

Yusoff, Kathryn. *A Billion Black Anthropocenes or None*. Minneapolis: University of Minnesota Press, 2018.

Yusoff, Kathryn. "The Inhumanities." *Annals of the American Association of Geographers* 111, no. 3 (2021): 663–76. https://doi.org/10.1080/24694452.2020.1814688.

Zimmerman, Rae. "Social Equity and Environmental Risk." *Risk Analysis* 13, no. 6 (1993): 649–66.

Zimring, Carl A. *Clean and White: A History of Environmental Racism in the United States*. New York: New York University Press, 2017.

Index

Emanuel African Methodist Episcopal
Church, 51–52, 189n65. *See also* Black
history
eminent domain, 6, 32, 39, 82, 123, 149
Engels, Frederich, 94
Environmental Protection Agency (EPA),
40–41, 43, 46, 77, 93, 95–96, 133–35,
187nn27–28, 188n31. *See also* Superfund
sites
ethnography, 10, 19–20, 22, 158, 163–64,
207n13

fabulation, 137, 139, 154, 162, 172; gender as,
141, 143, 153–54
Fanon, Frantz, 18, 31–32, 82, 176
femmeness, 103, 110–12, 116, 123–24; aesthetics, 23, 153; Black femmeness, 102,
104–5, 167
Fender Rhodes, 140, 152
Ferdinand, Malcom, 13, 62, 190n7
fertilizer, 40, 73, 76
Flint water crisis, 44, 97
flow, 58; flow control, 78, 193n52; and
racial capitalism, 156, 185n20; waste
flow, 8, 21, 24–29, 46, 63, 75, 92–96,
185n29, 193n52
Floyd, George, 45
flyness, 110, 112–13, 168, 174, 175; fly girl, 107
Fresh Kills, 24–25, 51, 184n2. *See also* landfill
fugitivity, 7, 20, 27, 58, 110, 123, 144; and
Blackness, 17, 22, 45, 56, 86, 123, 131; fugitive practices of Black women, 108–9,
112–13; fugitive living room, 37, 106, 139;
and gender, 140–41; geographies of, 52,
101, 190n68
fungibility, 34, 71, 98, 131, 190n77; of Black
flesh, 16; and capitalism, 161; and gender,
143; and property relations, 17; and terraforming, 33, 35; and toxic capture, 50

Gates Avenue, 114–15, 125, 171
Gaye, Marvin, 168–69
general strike, 192n12
genocide, 4, 9, 15, 74, 160, 200n67. *See also*
ecocide

genre of being, 13–14, 205n35; alternatives
to the human, 56, 86, 99, 102, 112, 119,
131, 155, 157; and gender, 142–43, 145, 151,
201n74; property as, 4–6, 162. *See also*
Black women; queerness
gentrification: in Brooklyn, 10, 22, 137–
39, 145, 154, 176; as dispossession, 33,
62, 99, 113–16, 142, 204n26; in Virginia,
30. *See also* Bedford-Stuyvesant
(Bed-Stuy); Darius; Hampton Roads;
scavenging
"Get Ur Freak On" (song), 110
Ghent, 30–31, 38–40
ghetto, 1–2, 117, 138–39, 159; ghetto trash,
3, 36
Gilmore, James, 26–27, 185nn18–19
Gilmore, Ruth Wilson, 19, 61, 191n17
Giuliani, Rudolph, 24–27, 51, 59, 146, 149,
179n8, 184n7
Goodie Mob, 157
Great Depression, 39, 125
Great Dismal Swamp, The, 40, 52, 66–67,
77, 80, 85–86, 88–90, 100, 190n68.
See also marronage; Nansemond
Great Migration, the, 27–28
Greenpeace, 93, 196n2, 196n4
Green Pines Apartments, 80–81

Hampton Roads, 9, 31–32, 39–40, 74, 155,
193n46; Hampton Roads Sanitation District, 194n66. *See also* gentrification
Harney, Stefano, 131
Harris, Cheryl, 13, 188n47. *See also* white
supremacy
Hartford, Connecticut, 2, 173
Hartman, Saidiya, 1, 20, 34, 84, 116, 122, 137,
204n15. *See also* fungibility
hip-hop, 35, 155, 186n18, 206n6; and deviant
Black aesthetics, 2, 103, 103, 109–10, 146,
149, 156–57, 160. *See also* Dirty South;
genre of being
Holiday Inn, 57, 65
Home Owners' Loan Corporation (HOLC),
39, 125–26
hood rat, 3, 36, 55–58, 176